"十三五"职业教育系列教材

电气技术专业英语
（第二版）

主编 朱一纶
编写 吴岱曦 吴 彪
主审 曹雪梅

中国电力出版社
CHINA ELECTRIC POWER PRESS

内 容 提 要

本书为"十三五"职业教育系列教材。

本书共分 15 个单元，主要内容包括电气工程简介、电路元器件、半导体器件、电路分析、交流电路、数字电路、传感器、电动机、控制电器、单片机、PLC、电力系统、自动化系统、电子 CAD、产品说明书等，每一单元分成课文和阅读材料两大部分，课文为基本内容，阅读材料则是相关内容的拓展，教师可以根据学生的能力和教学要求进行选择教学，每一单元附有基本的科技英语的翻译知识和一定量的练习。本书结合电气工程类专业的教学要求，参考选编了大量科技资料原文，并注意反映出最新的科技进展，内容生动，图文并茂，学生通过专业英语的学习可以拓展自己的专业词汇，提高阅读与专业相关的英文资料的能力，并可对电气专业知识作一回顾复习。

本书可作为高职高专院校电气、电子、自动控制等专业教材，也可作为有关工程技术类人员的参考书。

图书在版编目（CIP）数据

电气技术专业英语 / 朱一纶主编. —2 版. —北京：
中国电力出版社，2017.6（2023.1 重印）
"十三五"职业教育规划教材
ISBN 978-7-5198-0413-8

Ⅰ.①电… Ⅱ.①朱… Ⅲ.①电工技术—英语—职业教育—教材 Ⅳ.①TM

中国版本图书馆 CIP 数据核字（2017）第 054623 号

出版发行：中国电力出版社
地　　址：北京市东城区北京站西街 19 号（邮政编码 100005）
网　　址：http://www.cepp.sgcc.com.cn
责任编辑：冯宁宁（010-63412537）
责任校对：郝军燕
装帧设计：郝晓燕　赵姗姗
责任印制：吴　迪

印　　刷：北京雁林吉兆印刷有限公司
版　　次：2009 年 7 月第一版　2017 年 6 月第二版
印　　次：2023 年 1 月北京第十次印刷
开　　本：787 毫米×1092 毫米　16 开本
印　　张：13.25
字　　数：317 千字
定　　价：28.00 元

版权专有　侵权必究

本书如有印装质量问题，我社营销中心负责退换

  专业英语的教学目的是指导学生阅读与自己专业相关的英语书刊和文献资料，使学生以英语为工具，及时掌握最新的专业信息，在今后的工作中有更大的发展。

  本教材共分 15 个单元，适合于 32~48 学时的专业英语教学需要，每一单元分成课文和阅读材料两大部分，课文为基本内容，阅读材料则是相关内容的拓展，教师可以根据学生的能力和教学要求进行选择教学，每一单元附有基本的科技英语的翻译知识和一定量的练习。

  作为一本教材，本书具有以下的特点：

  （1）与专业结合紧密，希望学生通过专业英语的学习，可以拓展自己的专业词汇，提高阅读与专业相关的英文资料的能力，并可对电气专业知识作一回顾复习。

  （2）涉及的知识面比较广，选用了很多反映最新科技发展的课文资料和阅读资料，并注意选用不同题材的资料，以扩宽学生的知识面。

  （3）根据学生的英语基础选用原文资料并经适当改写，从浅入深，有利于逐步提高学生的英语阅读能力。

  （4）选用较多的插图，并制作与教材配套的电子教案，以加深学生的感性认识。

  本教材由南京金陵科技学院的朱一纶教授编写，吴岱曦参加了资料整理、文字录入和电子教案的制作工作，南京金陵科技学院的吴彪参加了校核等工作，在此表示感谢。

  本书由红河学院外国语学院曹雪梅主审。本书在编写过程中，得到许多同行的帮助，也引用、借鉴了相关专家的教材、著作，在此一并致谢。

  限于编者的学识水平与实践经验，书中难免存在不足之处，恳请读者和同行们批评指正。

  编者的电子邮箱：zhuyilun2002@163.com

编 者
2016 年 11 月

Contents

前言

Unit 1　Introduce to Electrical Engineering ……………………………………… 1
　1.1　Text ……………………………………………………………………………… 1
　1.2　Reading materials ……………………………………………………………… 4
　1.3　Knowledge about translation（科技英语的特点）………………………… 6
　1.4　Exercises ………………………………………………………………………… 7
　1.5　课文参考译文 …………………………………………………………………… 8
　1.6　阅读材料参考译文 ……………………………………………………………… 10

Unit 2　Basic Components ……………………………………………………………… 12
　2.1　Text ……………………………………………………………………………… 12
　2.2　Reading material ……………………………………………………………… 16
　2.3　Knowledge about translation（单词）……………………………………… 18
　2.4　Exercises ………………………………………………………………………… 20
　2.5　课文参考译文 …………………………………………………………………… 21
　2.6　阅读材料参考译文 ……………………………………………………………… 22

Unit 3　Semiconductor …………………………………………………………………… 24
　3.1　Text ……………………………………………………………………………… 24
　3.2　Reading material ……………………………………………………………… 29
　3.3　Knowledge about translation（非谓语动词Ⅰ）…………………………… 31
　3.4　Exercises ………………………………………………………………………… 32
　3.5　课文参考译文 …………………………………………………………………… 33
　3.6　阅读材料参考译文 ……………………………………………………………… 35

Unit 4　Analysis of an electric circuit ………………………………………………… 37
　4.1　Text ……………………………………………………………………………… 37
　4.2　Reading material ……………………………………………………………… 40
　4.3　Knowledge about translation（非谓语动词Ⅱ）…………………………… 42
　4.4　Exercises ………………………………………………………………………… 44
　4.5　课文参考译文 …………………………………………………………………… 45

4.6　阅读材料参考译文 ··· 47

Unit 5　Alternating current ··· 49
5.1　Text ·· 49
5.2　Reading material ··· 53
5.3　Knowledge about translation（被动语态）·· 55
5.4　Exercises ·· 57
5.5　课文参考译文 ··· 58
5.6　阅读材料参考译文 ··· 60

Unit 6　Digital System ··· 62
6.1　Text ·· 62
6.2　Reading materials ··· 67
6.3　Knowledge about translation（句子的连接Ⅰ）······································ 69
6.4　Exercises ·· 71
6.5　课文参考译文 ··· 72
6.6　阅读材料参考译文 ··· 74

Unit 7　Sensors ··· 77
7.1　Text ·· 77
7.2　Reading material ··· 80
7.3　Knowledge about translation（句子的连接Ⅱ）······································ 82
7.4　Exercise ·· 83
7.5　课文参考译文 ··· 84
7.6　阅读材料参考译文 ··· 86

Unit 8　Electric motor ··· 88
8.1　Text ·· 88
8.2　Reading material ··· 92
8.3　Knowledge about translation（分离现象）·· 94
8.4　Exercises ·· 96
8.5　课文参考译文 ··· 97
8.6　阅读材料参考译文 ··· 99

Unit 9　Motor Controller ··· 101
9.1　Text ·· 101
9.2　Reading materials ··· 104
9.3　Knowledge about translation（省略和插入语）···································· 108
9.4　Exercises ·· 110

9.5	课文参考译文	111
9.6	阅读材料参考译文	112

Unit 10 Power System ... 115
10.1	Text	115
10.2	Reading materials	119
10.3	Knowledge about translation（It 的用法）	122
10.4	Exercises	124
10.5	课文参考译文	125
10.6	阅读材料参考译文	127

Unit 11 Microcontroller ... 130
11.1	Text	130
11.2	Reading materials	134
11.3	Knowledge about translation（That 的用法）	137
11.4	Exercises	138
11.5	课文参考译文	139
11.6	阅读材料参考译文	141

Unit 12 Programmable Controller ... 144
12.1	Text	144
12.2	Reading materials	149
12.3	Knowledge about translation（Which 的用法）	152
12.4	Exercises	153
12.5	课文参考译文	154
12.6	阅读材料参考译文	157

Unit 13 Automation ... 159
13.1	Text	159
13.2	Reading materials	164
13.3	Knowledge about translation（倒装）	166
13.4	Exercises	168
13.5	课文参考译文	169
13.6	阅读材料参考译文	171

Unit 14 Electronic Design Automation ... 173
14.1	Text	173
14.2	Reading materials	177
14.3	Knowledge about translation（否定的表示）	179

	14.4	Exercises ··· 181

14.4 Exercises ··· 181
14.5 课文参考译文 ·· 182
14.6 阅读材料参考译文 ··· 184

Unit 15　User Manual ·· 187
15.1 Text ·· 187
15.2 Reading materials ·· 192
15.3 Knowledge about translation（用户说明书）···································· 195
15.4 Exercises ·· 196
15.5 课文参考译文 ·· 198
15.6 阅读材料参考译文 ··· 200

参考文献 ·· 203
参考资料 ·· 203

Unit 1 Introduce to Electrical Engineering

1.1 Text

Electricity is the most common form of energy. Electricity is used for various applications such as lighting, transportation, cooking, communication, production of various goods in factories and much more.

1.1.1 What is electrical engineering

Electrical engineering, sometimes referred to as electrical and electronic engineering, is a field of engineering that deals with the study and application of electricity, electronics and electromagnetism (Fig1.1). The field first became an identifiable occupation in the late nineteenth century after commercialization of the electric telegraph and electrical power supply. Electrical engineering has now subdivided into a wide range of subfields including electronics, digital computers, power engineering, telecommunications, control systems, radio-frequency engineering, signal processing, instrumentation, and microelectronics.

Electrical engineering may or may not encompass electronic engineering. Where a distinction is made, usually outside of the United States, electrical engineering is considered to deal with the problems associated with large-scale electrical systems such as power transmission and motor control, whereas electronic engineering deals with the study of small-scale electronic systems including computers and integrated circuits. Alternatively, electrical engineers are usually concerned with using electricity to transmit energy, while electronic engineers are concerned with using electricity to transmit information.

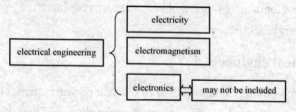

Fig 1.1 range of electrical engineering

1.1.2 History

Electricity has been a subject of scientific interest since at least the early 17th century. However, it was not until the 19th century that research into the subject started to intensify. Notable developments in this century include the work of George Simon Ohm[1], who in 1827 quantified the relationship between the electric current and potential difference in a conductor, Michael Faraday[2], the discoverer of electromagnetic induction in 1831, and James Clerk Maxwell[3], who in 1873 published a unified theory of electricity and magnetism in his treatise Electricity and Magnetism.

During this period, the work concerning electrical engineering increased dramatically. In 1882, Thomas Alva Edison[4] [Fig 1.2(a)] switched on the world's first large-scale electrical supply network that provided 110 volts direct current (DC) to fifty-nine customers in lower Manhattan. In the same time, alternating current, with its ability to transmit power more efficiently over long distances via the use of transformers, developed rapidly in the 1880s. In 1887, Nikola Tesla[5] [Fig 1.2(b)] filed a number of patents related to a competing form of power distribution known as alternating current (AC). In the following years a bitter rivalry known as the "War of Currents", took place over the preferred method of distribution. AC eventually replaced DC for generation and power distribution, enormously extending the range and improving the safety and efficiency of power distribution.

(a)　　　　　　　　　　　　　　(b)

Fig 1.2　two Famous Scientists
(a) **Thomas Alva Edison**　built the world's first large-scale electrical supply network;
(b) **Nikola Tesla**　made long-distance electrical transmission networks possible

The efforts of the two did much to further electrical engineering—Tesla's work on induction motors and polyphase systems influenced the field for years to come, while Edison's work on telegraphy and his development of the stock ticker proved lucrative for his company, which ultimately became General Electric.

1.1.3　What electrical engineers do?

From the Global Positioning System (GPS) to electric power system, electrical engineers have contributed to the development of a wide range of technologies. They design, develop, test and supervise the deployment of electrical systems and electronic devices. For example, they may work on the design of telecommunication systems, the operation of electric power stations, the lighting and wiring of buildings, the design of household appliances or the electrical control of industrial machinery.

Today most engineering work involves the use of computers and it is common place to use computer-aided design programs when designing electrical systems. Nevertheless, the ability to sketch ideas is still invaluable for quickly communicating with others.

Although most electrical engineers will understand basic circuit theory (that is the interactions of elements such as resistors, capacitors, diodes, transistors and inductors in a circuit), the theories employed by engineers generally depend upon the work they do. For example, quantum mechanics and solid state physics might be relevant to an engineer working on VLSI (the design of integrated circuits), but are largely irrelevant to engineers working with macroscopic electrical systems. Even circuit theory may not be relevant to a person designing telecommunication systems that use off-the-shelf components. Perhaps the most important technical skills for electrical engineers are reflected in university programs, which emphasize strong numerical skills, computer literacy and the ability to understand the technical language and concepts that relate to electrical engineering.

The workplaces of electrical engineers are just as varied as the types of work they do. Electrical engineers may be found in the lab environment of a fabrication plant, the offices of a consulting firm or on site at a mine. During their working life, electrical engineers may find themselves supervising a wide range of individuals including scientists, electricians, computer programmers and other engineers.

 Review

(1) Electrical engineering is a field of engineering that deals with the study and application of electricity, electronics and electromagnetism.
(2) Eelectrical engineers design, develop, test and supervise the deployment of electrical systems and electronic devices.
(3) The workplaces of electrical engineers are just as varied as the types of work they do.

📖 Notes to the text

[1] **George Simon Ohm** (1789—1854) 德国物理学家，欧姆定律就是以他命名的。
[2] **Michael Faraday** (1791—1867) 英国物理学家和化学家，毕生致力于电磁学和电化学的研究，法拉第定律就是以他命名的。
[3] **James Clerk Maxwell** (1831—1879) 苏格兰数学家和理论物理学家，他最杰出的成就是建立了电磁场方程，麦克斯韦方程就是以他命名的。
[4] **Thomas Alva Edison** (1847—1931) 美国的发明家爱迪生，他一生获得很多发明专利，这里提到他建立了第一个小规模的直流供电系统。
[5] **Nikola Tesla** (1856—1943) 美国发明家，机械师和电气工程师，在电磁学、交流电等方面有很多研究，国际单位制中电磁场磁感应强度 B 的单位"特斯拉，T"就是以他的名字命名的。

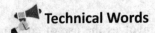 **Technical Words**

circuit ['sɜːkit] n. 电路,一圈,周游,巡回 n. 线路(系统,环行) vt. 接成电路(绕……环行)
commercial [kə'mɜːʃ(ə)l] adj. 商业的,贸易的 commercialization n. 商业化,商品化
deploy [di'plɔi] v. 展开,配置 deployment n. 配置
distinction [di'stiŋ(k)ʃ(ə)n] n. 区别,差别,级别,特性
electricity [i'lektrisəti] n. 电流,电,电学
electromagnetism [i,lektrəʊ'mægnitiz(ə)m] n. 电磁,电磁学
emphasize ['emfəsaiz] vt. 强调,着重
fabrication [fæbri'keiʃ(ə)n] n. 制造,构成,装配工
identifiable [aidenti'fæiəb(ə)l] adj. 可以确认的
integrated ['intigreitid] adj. 综合的,完整的,集成的
invaluable [in'væljʊ(ə)b(ə)l] adj. 无价的,价值无法衡量的
literacy ['lit(ə)rəsi] n. 有文化,有教养,这里指计算机方面的能力
macroscopic [,mækrə(ʊ)'skɔpik] adj. 肉眼可见的,宏观的
magnetism ['mægnitiz(ə)m] n. 磁,磁力,吸引力,磁学
motor ['məʊtə] n. 电动机

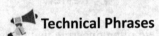 **Technical Phrases**

electrical engineering	电气工程
electronic engineering	电子工程
AC (alternating current)	交流电(流)
DC (direct current)	直流电(流)
Global Positioning System (GPS)	全球定位系统
electric power system	电力系统
power distribution	电力分配(配电)
polyphase system	多相系统(如三相交流电系统)

1.2 Reading materials

Electrical engineering has many sub-disciplines, the most popular of which are listed below. Although there are electrical engineers who focus exclusively on one of these sub-disciplines, many deal with a combination of them. Sometimes certain fields, such as electronic engineering and computer engineering, are considered separate disciplines in their own right.

1.2.1 Power engineering

Power engineering deals with the generation, transmission and distribution of electricity as well as the design of a range of related devices. These include transformers, electric generators, electric motors, high voltage engineering and power electronics. In many regions of the world, governments maintain an electrical network called a power grid that connects a variety of generators together with users of their energy. Users purchase electrical energy from the grid, avoiding the costly exercise of having to generate their own. Power engineers may work on the design and maintenance of the power grid as well as the power systems that connect to it. Such systems are called on-grid power systems and may supply the grid with additional power, draw power from the grid or do both. Power engineers may also work on systems that do not connect to the grid, called off-grid power systems.

1.2.2 Control engineering

Control engineering focuses on the modeling of a diverse range of dynamic systems and the design of controllers that will cause these systems to behave in the desired manner. To implement such controllers electrical engineers may use electrical circuits, digital signal processors, microcontrollers and PLCs (Programmable Logic Controllers). Control engineering has a wide range of applications from the flight and propulsion systems of commercial airliners to the cruise control present in many modern automobiles. It also plays an important role in industrial automation.

In most of the cases, control engineers utilize feedback when designing control systems. For example, in an automobile with cruise control the vehicle's speed is continuously monitored and fed back to the system which adjusts the motor's torque accordingly. Where there is regular feedback, control theory can be used to determine how the system responds to such feedback. In practically all such systems stability is important and control theory can help ensure stability is achieved.

1.2.3 Instrumentation engineering

Instrumentation engineering deals with the design of devices to measure physical quantities such as pressure, flow and temperature. Instrumentation is an electrical device placed in the field to provide measurement and/or control capabilities for the system. The design of such instrumentation requires a good understanding of physics that often extends beyond electromagnetic theory. For example, flight instruments measure variables such as wind speed and altitude to enable pilots the control of aircraft analytically.

Control instrumentation includes devices such as solenoids, Electrically Operated Valves, breakers, relays, etc. These devices are able to change a field parameter, and provide remote control capabilities.

Transmitters are devices which produce an analog signal, usually in the form of a 4-20mA electrical current signal, although many other options are possible using voltage, frequency, or pressure. This signal can be used to directly control other instruments, or sent to a PLC, DCS system or other type of computerized controller, where it can be interpreted into readable values, or used to control other devices and processes in the system.

Instrumentation plays a significant role in both gathering information from the field and changing the field parameters, and as such are a key part of control loops.

Instrumentation engineering is the engineering specialization focused on the principle and operation of measuring instruments which are used in design and configuration of automated systems. They typically work for industries with automated processes, such as chemical or manufacturing plants, with the goal of improving system productivity, reliability, safety, optimization and stability.

1.3　Knowledge about translation（科技英语的特点）

科技英语一般都涉及各专业领域的新技术、新发展。针对不同的对象，格式也稍有不同，例如作为科普读物，专业词汇一般用得就比较少，用到时也尽可能做详细解释；作为专业论文，则会用到很多专业词汇，叙述客观、正确；作为专业资料，则比较精练，但总体看，有四个特点：

1. 专业词汇多

在科技英语阅读中，单词有三个方面要注意：

首先，单词有不少是多性词，即既是名词，又可作动词、形容词、介词或副词，字形无殊，功能各异，阅读时也很容易造成曲解。

如：light

用作名词：safety light　　　　安全灯
形容词：light coating　　　　薄涂层
动词：light up the lamp　　　点灯
副词：travel light　　　　　　轻装旅行

其次有些英语单词在不同的专业中使用时有特定的含义，翻译时要根据上下文的意思选取词意。

如：power

数学：乘方、幂
物理：电路：功率
电力系统：电力

第三是缩写词用得比较多，但一般在首次提到时都会给出缩写词的全文，以后就直接用缩写词了。

如：Global Positioning System(GPS)——全球定位系统
　　alternating current（AC）——交流电

2. 被动语态多

科技英语中在描述客观事物时，常采用被动语态，通过把所论述的过程或事实放在句子的首位，可以突出其重要性。

Electrical engineering is considered to deal with the problems associated with large-scale electrical systems.

通常认为电气工程是处理与大规模电气系统相关的问题。

The values of the resistor are calculated from the color of the bands.

可以根据色码条的颜色求出电阻的值。

在翻译中可以根据中文的习惯，不一定要译出被动语态。

3. 非谓语动词多

英语语法中每个简单句中只能用一个谓语动词，如果一个句子中有几个动作就必须选出主要动作当谓语，而将其余动作用非谓语动词形式（V-ing，V-ed，to V 三种形式），才能符合英语的语法要求。

Electrical engineers are usually concerned with using electricity to transmit energy.

电气工程师通常关心电能的传输问题。

这里主要动词是 are concerned (被动态)，而 using，to transmit 都是非谓语动词形式描述动作。

非谓语动词也常用作定语等。

They typically work for industries with automated processes, such as chemical or manufacturing plant.

仪器工程的典型工作环境是带有自动化过程的工业，如化学工厂或制造工厂。

这里非谓语动词 automated 作形容词，形容 processes，非谓语动词 manufacturing 形容 plant。

4. 复杂长句多

科技文章要求叙述准确，用词严谨，因此一句话里常常包含多个分句，这种句子成分复杂且很长的句子是阅读科技英语的第一个难点，阅读翻译时要按汉语习惯来加以分析，把句子断开，以短代长，化难为易。

Electrical engineering, sometimes referred to as electrical and electronic engineering, is a field of engineering that deals with the study and application of electricity, electronics and electromagnetism.

电气工程，有时称为电气和电子工程，是研究电、电子和电磁效应及其应用的一个工程领域。

这里，要特别注意标点符号和一些连接词，如上例中的 that 可以帮助我们断开长句。在科技英语中，要注意 it, that, which 等词的指代，有时要结合自己的专业知识来翻译。

1.4　Exercises

1. Put the Phrases into English (将下列词组译成英语)

(1) 电子工程

(2) 计算机辅助设计

(3) 小规模电子系统

(4) 控制系统

(5) 电磁感应

(6) 电力分配

(7) 电气控制

(8) 交流电

(9) 感应（异步）电动机

(10) 直流电

2. Put the Phrases into Chinese (将下列词组译成中文)

(1) electrical power supply

(2) referred to as

(3) electrical engineering

(4) deal with the problems

(5) transmit energy

(6) in the lab environment

(7) interaction of elements

(8) solid state physics

(9) be largely irrelevant to

(10) polyphase system

3. Sentence Translation (将下列句子译成中文)

(1) Electrical engineering may or may not encompass electronic engineering.

(2) Electricity has been a subject of scientific interest since at least the early 17th century.

(3) James Clerk Maxwell, who in 1873 published a unified theory of electricity and magnetism in his treatise Electricity and Magnetism.

(4) Electrical engineers have contributed to the development of a wide range of technologies.

(5) The workplaces of electrical engineers are just as varied as the types of work they do.

4. Translation (翻译)

Computer engineering deals with the design of computers and computer systems. This may involve the design of new hardware, the design of PDAs or the use of computers to control an industrial plant. Computer engineers may also work on a system's software. However, the design of complex software systems is often the domain of software engineering, which is usually considered a separate discipline.

1.5 课文参考译文

电力是能量的最常见形式。电力被用于各个领域，例如照明、交通、烹饪、通信、在工厂生产各种产品等。

1.5.1　什么是电气工程

电气工程，有时称为电气和电子工程，是研究电气、电子和电磁效应及其应用的一个工程领域（图1.1）。在19世纪后期实现了电信和电能输送以后，这个领域形成一个行业领域。电气工程目前已细分成包括电子、数字计算机、电力工程、通信、控制系统、无线电频率工程、信号处理、仪器仪表和微电子等一系列的子专业领域。

在美国电气工程是包含电子工程的，而其他国家电气工程一般不包括电子工程，通常认为电气工程是处理如电能传输、电动机控制等大规模电气系统相关问题的（中国有时称为强电）。而电子工程是则处理计算机、集成电路等小规模电子系统的问题（中国有时称为弱电）。相应的，电气工程师通常关心电能的传输问题，而电子工程师则关心电信号的传输问题。

1.5.2　历史

17世纪早期科学家们就开始了对电的研究，但直到19世纪对电的研究才有了较深入的进展，这个世纪中值得注意的工作进展包括乔治·西蒙·欧姆，他在1827年确立了导体的（两端的）电位差（电压）和（流过导体的）电流的关系；米歇尔·法拉第，他在1831年发现了电磁感应（定律）；詹姆斯·克拉克·麦克斯韦，他在1873年发表了"电流和磁场"论文，提出了电磁场统一的理论（Maxwell方程）。

在这个阶级，涉及电气工程的工作进展十分快。1882年，托马斯·阿尔瓦·爱迪生［图1.2（a）］建立了世界上第一个大规模电力传输网，为曼哈顿地区的59位用户输送110V的直流电源。同时，交流电，因为它通过使用变压器具有更有效的长距离传输电能的能力，在19世纪80年代得到迅速发展。1887年，尼古拉·特斯拉［图1.2（b）］申请了很多有关交流电配电（电力分配）方面的有竞争力的专利。接下来的几年中，开始了一场关于输配电方法上的激烈的竞争称为"电流的战争"，最终交流电取代了直流电，成为发电和输电的主要方式，并且扩展了输电范围，在安全性和提高输电效率方面也有了很大的提高。

这两个人的研究对电气工程的发展都有很大贡献，特斯拉关于感应电动机和多相输电系统的研究到今天还影响着电气工程领域，而爱迪生关于电报和证券报价机等的发明为他的公司带来了很大的利益，他的公司最终成为（美国）通用电气公司。

1.5.3　电气工程师做什么工作

从全球定位系统到电力系统，电气工程师对多个技术发展领域做出了很多贡献。他们在电气系统和电子器件方面进行设计、开发、测试和监控工作，例如，他们可以做通信系统的设计工作，也可以做发电站的运行管理、建筑物的照明布线、家用电器的设计或工业设备的电气控制等各种工作。

现代的电气工程师的工作要用到计算机，通常是在设计电气系统时用计算机辅助设计编程，但仍要具有用草图（或基本概念）与别人交流设计思想的能力。

虽然大部分电气工程师都学过基本电路理论（即电路中电阻、电容、二极管、三极管和电感等基本元件及其元件的相互关系），根据工程师们做的工作不同，他们对这些理论的应用要求也不同。例如，一个设计集成电路的工程师可能要用到量子力学和固态电子学，但对一个宏观电气系统设计的工程师来说,这些内容(量子力学和固态电子学)与他们无关。对一个设计远程通信系统的工程师而言,甚至电路理论也很少用到,他们只要用现成的电器元件。也许对一个电气工程师来说,最重要的技能可以用大学的课程来反映,大学课程强调的是数字化技术,计算机知识和掌握与电气工程相关的技术语言和概念的能力。

由于电气工程师的工作性质不同,他们的工作场合也发生变化,他们可以在制造工厂的实验环境工作,也可以在咨询公司办公室中或矿山的现场工作。在电气工程师的工作生涯中,可能有很多不同的个人经历,包括科学家、电工、计算机程序员和其他工程师。

1.6 阅读材料参考译文

电气工程有很多子学科,以下列出其中最常开设的一些子学科。虽然有些电气工程师是只偏重于其中一个子学科,也有很多电气工程师是组合型的(即对多个子学科都有研究)。有时一些领域,例如,电子工程和计算机工程可以看成是各自独立的学科。

1.6.1 电力工程

电力工程涉及电能的产生、输送、分配及相关设备的设计。这里包括变压器、发电机、电动机、高压工程和电力电子。世界上很多地方,是政府建设维护电力网,通过电力网把许多发电机与用户相连接,用户购买电网输送的电能,不必自己发电,可以降低成本。电力系统工程师的工作可以是设计和维护电力网和连接在电力网上电力系统(设备)。这些系统称为电网上的电力系统,是指给电网供应电能,获取电能或两者都有的电力系统。电力系统工程师的工作也可以是设计和维护不与电网相连接的供电、用电设备,称为非电网系统(设备)。

1.6.2 控制工程

控制工程是通过对各种各样动力系统的建模和控制器的设计,使得这些动力系统能按要求的方式工作。电气工程师可以用电路、数字信号处理器、微控制器(单片机)和PLC(可编程控制器)实现这样的控制。控制工程有很广的应用范围,从飞机的推进系统及飞行过程的控制到现代许多车辆行驶的控制,在工业自动化生产中也起着重要的作用。

在很多控制系统设计中,控制工程师利用反馈,例如,在带有速度控制的汽车中,控制系统不断检测车辆的速度并反馈到系统中用以调节电动机的转矩。在一般含反馈的系统中,是用控制理论来求出系统对这些反馈应采取什么响应。实际上所有这些系统的稳定性是很重要的,控制理论可以保证使系统实现稳定。

1.6.3 仪器工程

仪器工程涉及设计测量物理量(如压力、流量和温度)的仪器。仪器是指工业现场用于测量和控制系统的电气设备。要设计这些仪器不仅需要很好地掌握电磁理论,还需要掌握其他的物理学知识,例如,飞行仪器要测量风速和高度等变量,飞行员根据这些数据控制飞机的飞行。

　　控制仪器包括如螺旋管、电控阀、断路器、继电器等，这些器件能用来改变现场参数，提供遥控的手段。

　　变送器是输出模拟信号仪器，一般输出 4~20mA 的电流信号，也可以用电压、频率或压力作为输出信号。这些输出信号一般可以直接控制其他的仪器，或送给一个可编程控制器（PLC）、分散集成系统（DCS）或其他类型的计算机控制器，计算机可把这些输出信号译成可读的数值或用于控制系统的其他器件和工作过程。

　　仪器在获取现场信息和改变现场参数这两个方面都起着重要的作用。因此是控制环节的一个关键部分。

　　仪器工程是研究测量仪器的原理和使用的工程专业，这些仪器用在自动化系统的设计和设置中。仪器工程的典型工作环境是带有自动化过程的工业，如化学工厂或制造工厂，目标是改进系统的产量，提高可靠性、安全性、优化和稳定性。

Unit 2　Basic Components

2.1　Text

Basic components usually are resistors, capacitors, inductor, they are not always as simple as they may appear at first look. We need to know something about them before starting on even the simplest of projects.

2.1.1　Resistor

The first and most common electronic component is the resistor. There is virtually no working circuit doesn't use them.

Suppose that some material is connected to the terminals of an ideal voltage source as shown in Fig 2.1(a), if the resulting current $i(t)$ is always directly proportional to the voltage for any function $v(t)$, then the material is called a linear resistor, or resistor for short.

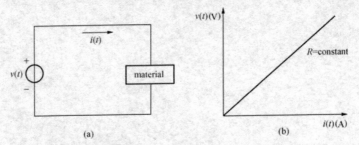

Fig 2.1　the relationship of voltage and current
(a) connected circuit; (b) $i(t)$ is directly proportional to $v(t)$

Since voltage and current are directly proportional for a resistor, there exists a proportionality constant R, called resistance.

The amount of current flowing in a resistor is directly proportional to the voltage across it and inversely proportional to the resistance of the resistor. This is Ohm law and can be expressed as a formula: $v(t)=Ri(t)$.

The unit of resistance (volts per ampere) is referred to as Ohms, and is denoted by the capital Greek letter omega, Ω. A plot of voltage versus current for a (linear) resistor is given in Fig 2.1(b).

Resistors are used to limit current flowing to a device, thereby preventing it from burning out, as voltage dividers to reduce voltage for other device, as transistor biasing circuits, and to serve as circuit loads. Fig 2.2 (a) shows the different resistor symbols that are used in circuit diagrams. The rectangular box is used throughout Europe, while the zig-zag line is more common in Japan and the USA.

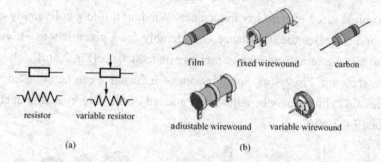

Fig 2.2 resistors
(a) symbols; (b) resistors in common use

2.1.2 Capacitor

A capacitor is an electronic device for temporarily store electrical energy. Capacitors can be found in almost any complex electronic circuit. There are many different types of capacitor but they all work in essentially the same way. A simplified view of a capacitor is a pair of metal plates separated by a gap in which there is an insulating material known as the dielectric. This simplified capacitor is also chosen as the electronic circuit symbol for a capacitor is a pair of parallel plates as shown in Fig 2.3(a). Some capacitors' capacitance can be adjusted, so they are variable capacitors [Fig 2.3(b)].

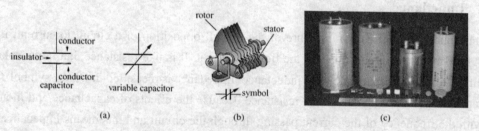

Fig 2.3 capacitors
(a) symbols; (b) air variable capacitor; (c)various capacitors

If voltage is applied to the capacitor terminals, charge flows in and collects on one plate. Meanwhile, current flows out of the other terminal, and a charge of opposite polarity collects on the other plate.

The magnitude of the net charge Q on one plate is proportional to the applied voltage V. Thus, we have $Q=CV$, in which C is the capacitance.

A capacitor will block dc current, but appears to pass ac current by charging and discharging. It develops an ac resistance, known as capacitive reactance, which is affected by the capacitance and ac frequency. The formula for capacitive reactance is $X_C=1/(2\pi f_C C)$, with units of Ohms Ω.

2.1.3 Inductor

An inductor is an electrical device, which can temporarily store electromagnetic energy in the field about it as long as current is flowing through it. An inductor is most commonly a coil, but in reality, even a straight piece of wire has inductance. Winding it into a coil simply concentrates the magnetic field, and increases the inductance considerably for a given length of wire. The inductor coil may have an air core or an iron core to increase its inductance[Fig 2.4(c)]. The circuit symbols for inductors are shown in Fig 2.4(a). Some inductors' inductance can be adjusted; a powered iron core in the shape of a cylinder may be adjusted in and out of the core in such inductors, so they are variable inductors[Fig 2.4(b)].

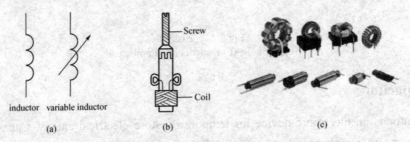

Fig 2.4 inductors
(a) symbols; (b) variable inductor; (c) various inductors

An inductor tends to oppose a change in electrical current, it has no resistance to dc current but has an ac resistance to ac current, known as inductive reactance, this inductive reactance is affected by inductance and the ac frequency and is given by the formula $X_L = 2\pi f_L L$, with units of Ohms.

2.1.4 Impedance

Impedance (symbol Z) is a measure of the overall opposition of a circuit to current, in other words: how much the circuit impedes the flow of current. It is like resistance, but it also takes into account the effects of capacitance and inductance. Impedance is measured in Ohms, symbol Ω.

Impedance is more complex than resistance because the effects of capacitance and inductance vary with the frequency of the current passing through the circuit and this means impedance varies with frequency. The effect of resistance is constant regardless of frequency.

The term 'impedance' is often used (quite correctly) for simple circuits which have no capacitance or inductance-for example to refer to their 'input impedance' or 'output impedance'. This can seem confusing if you are learning electronics, but for these simple circuits you can assume that it is just another word for resistance.

Four electrical quantities determine the impedance (Z) of a circuit: resistance (R), capacitance (C), inductance (L) and frequency (f).

Impedance can be split into two parts: $Z = R + jX$

Resistance R (the part which is constant regardless of frequency)

Reactance X (the part which varies with frequency due to capacitance and inductance)

Review

(1) The amount of current flowing in a resistor is directly proportional to the voltage across it.
(2) A capacitor is an electronic device for temporarily store electrical energy.
(3) An inductor is an electrical device, which can temporarily store electromagnetic energy in the field about it.
(4) Impedance (symbol Z) is a measure of the overall opposition of a circuit to current.

Technical Words

capacitance [kəˈpæsit(ə)ns] *n.* 电容量
capacitor [kəˈpæsɪtə] *n.* (=capacitator) 电容器
charge [tʃɑːdʒ] *n.* 负荷，电荷，费用，主管，掌管，充电，充气，装料 *v.* 装满，控诉，责令，告诫，指示，加罪于，冲锋，收费
core [kɔː] *n.* 核心，中心
current [ˈkʌr(ə)nt] *n.* 电流
cylinder [ˈsilində] *n.* 圆筒，圆柱体，汽缸，柱面
dielectric [ˌdaiiˈlektrɪk] *n.* 电介质，绝缘体 *adj.* 非传导性的
formula [ˈfɔːmjʊlə] *n.* 公式，规则，客套语
impedance [imˈpiːd(ə)ns] 阻抗（包括电阻与电抗）
inductance [inˈdʌkt(ə)ns] *n.* 电感量，电感值
inductor [inˈdʌktə] *n.* 诱导物，感应器，电感器
insulating [ˈinsəletiŋ] *adj.* 绝缘的
inverse [ˈinvɜːs; inˈvɜːs] *adj.* 相反的 inversely *adv.* 相反地，倒转地
linear [ˈliniə] *adj.* 直线的，线性的
proportionality [prəˌpɔːʃəˈnæləti] 比例（性），均衡（性）
reactance [riˈækt(ə)ns] *n.* 电抗
resistance [riˈzist(ə)ns] *n.* [电]电阻值，中文中有时也简称电阻
resistor [riˈzistə] *n.* [电]电阻器，中文中有时简称电阻
temporary [ˈtemp(ə)rəri] *adj.* 暂时的；临时的；短暂的；一时的；temporarily *adv.*
terminal [ˈtɜːmin(ə)l] *n.* 终点站，终端，接线端
versus [ˈvɜːsəs] *prep.* 对（指诉讼，比赛等中），与……相对，常缩写 vs.
voltage [ˈvəʊltidʒ] *n.* [电工]电压，伏特数 *n.* 电压（电位差）
zig-zag [zigˈzæg] *n.* Z字形，锯齿形，蜿蜒曲折 *v.* 作Z字形行进 *adj.* Z字形的

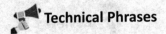 **Technical Phrases**

be connected to	与……相连接
be directly proportional to	正比于……
be split into	分成……
be affected by	受……影响
transistor biasing circuits	晶体管偏置电路
capital Greek letter	大写希腊字母

2.2 Reading material

2.2.1 Resistor Color Codes

Low power (≤2W) resistors are coded with colors to identify their value and tolerance. This comes in two variants-4 band and 5 band. The 4 band code is most common with 5% and 10% tolerance, and the 5 band code is used with 1% and better.

The values of the resistor are calculated from the color of the bands (see Fig 2.5). The values of the colors are shown in Table 2.1. For a 4-band resistor, the first band is the tens values. The second band gives the units; the third band is a multiplying factor, the factor being 10's band value. The fourth band gives the tolerance of the resistor. Some times no fourth band implies a tolerance of ±20%, a silver band means the resistor has a tolerance of ±10% and a gold band has the closest tolerance of ± 5%.

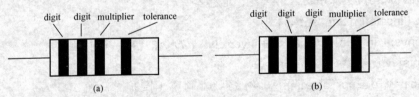

Fig 2.5 resistor color codes
(a) 4-band code; (b) 5-band code

Table 2.1   **The values of the colors**

Color	Digit	Multiplier	Tolerance(%)	Color	Digit	Multiplier	Tolerance(%)
Black	0	$10^0(1)$		Violet	7	10^7	0.1
Brown	1	10^1	1	Grey	8	10^8	
Red	2	10^2	2	White	9	10^9	
Orange	3	10^3		Gold		10^{-1}	5
Yellow	4	10^4		Silver		10^{-2}	10
Green	5	10^5	0.5	(none)			20
Blue	6	10^6	0.25				

For a 5-band resistor, the first band is the hundreds values. The second band gives the tens and the third band gives units. The forth band is a multiplying factor, the factor being 10's band value. The colors brown, red, green, blue, and violet are used as tolerance codes on 5-band resistors only. All 5-band resistors use a colored tolerance band.

Fig 2.6 gives some examples of color code resistor. Resistor in Fig 2.6(a) colored Brown-Green-Grey-Silver-Red would be 1.58 Ω with a tolerance of ± 2% and resistor in Fig 2.6(b) colored Yellow-Violet-Orange-Gold would be 47 kΩ with a tolerance of ± 5%.

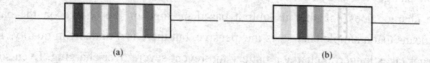

Fig 2.6　examples of color code resistor
(a) Brown-Green-Grey-Silver-Red; (b)Yellow-Violet-Orange-Gold

2.2.2　Domestic power plugs and sockets

In most countries, household power is single-phase electric power, in which a single phase conductor brings alternating current into a house, and a neutral returns it to the power supply.

Domestic power plugs and sockets are devices that connect the home appliances and portable light fixtures commonly used in homes to the commercial power supply so that electric power can flow to them. Electrical plugs and sockets differ in voltage and current rating, shape, size and type of connectors. The types used in each country are set by national standards(Fig 2.7),Many plugs and sockets include a third contact used for a protective earth ground, which only carries current in case of a fault in the connected equipment.

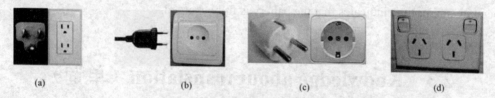

Fig 2.7　Domestic power various plugs & sockets
(a) North American 15 A/125 V grounded; (b) Japanese 15 A/100 V;
(c) German 16 A/250 V earthed; (d) Australian/New Zealand & Chinese10 A/240 V

Power plugs are male electrical connectors that fit into female electrical sockets. They have contacts that are pins or blades that connect mechanically and electrically to holes or slots in the socket. Plugs usually have a phase or hot contact, a neutral contact, and an optional earth or Ground contact. Many plugs make no distinction between the live and neutral contacts. The contacts may be steel or brass, either zinc, tin or nickel plated.

Power sockets are female electrical connectors that have slots or holes which accept the pins or blades of power plugs inserted into them and deliver electricity to the plugs. Sockets are usually designed to reject any plug which is not built to the same electrical standard. Some sockets have one or more holes that connect to pins on the plug.

2.2.3 Battery charger

A battery charger is a device used to put energy into a secondary cell or (rechargeable) battery by forcing an electric current through it.

The charge current depends upon the technology and capacity of the battery being charged. For example, the current that should be applied to recharge a 12V car battery will be very different from the current for a mobile phone battery.

A simple charger works by connecting a constant DC power source to the battery being charged. The simple charger does not alter its output based on time or the charge on the battery. This simplicity means that a simple charger is inexpensive, but there is a tradeoff in quality. Typically, a simple charger takes longer to charge a battery to prevent severe over-charging. Even so, a battery left in a simple charger for too long will be weakened or destroyed due to over-charging.

A trickle charger is a kind of simple charger that charges the battery slowly, at the self-discharge rate. A trickle charger is the slowest kind of battery charger. Leaving a battery in a trickle charger keeps the battery "topped up" but never over-charges.

The output of a timer charger is terminated after a pre-determined time. Timer chargers were the most common type for high-capacity Ni-Cd cells in the late 1990s. Often a timer charger and set of batteries could be bought as a bundle and the charger time was set to suit those batteries. If batteries of lower capacity were charged then they would be overcharged, and if batteries of higher capacity were charged they would be only partly charged.

An intelligent charger is that it's output current depends upon the battery's state. An intelligent charger may monitor the battery's voltage, temperature and/or time under charge to determine the optimum charge current at that instant. Charging is terminated when a combination of the voltage, temperature and/or time indicates that the battery is fully charged.

2.3 Knowledge about translation（单词）

1. 单词辨析

英语的单词大多数都是多义词，应注意准确选定英语单词在句子中的含义。根据文章涉及的专业内容来确定其含义是一种有效的方法。

（1）Charge：

service charge	服务费
in charge of	负责……
furnaces charge	炉料
induced charge	感应电荷

（2）Develop：

economic development	经济发展

intellectual development　　智力开发
series development　　级数展开
chemical development　　化学显影（冲胶卷）

在阅读电子技术专业文献时，注意扩大自己的专业词汇，许多平时很熟悉的单词在电子专业文献中可能有其特定的含义。

2. 单词的词义、词性

（1）英语语法有它的词汇构成方法，在阅读中注意辨别和记忆。

例如：最常用的电阻、电容、电感，注意分辨下列各词的含义和构成特点：

device		*value*		*adj.*	
resistor	电阻（器）	resistance	电阻（值）	resistive circuit	阻性电路
capacitor	电容（器）	capacitance	电容（值）	capacitive reactance	容抗
inductor	电感（器）	inductance	电感（值）	inductive reactance	感抗

中文往往把电阻器和电阻值均称为电阻，这样的情况在专业文章里也要注意。

（2）前缀与后缀。

英语的很多单词，通过加上前缀或后缀就可以改变其词性和含义，了解并掌握这一点对阅读科技英语很有帮助。这里举几个科技英语中常用的例子：

例如：

动词加 able 后缀构成形容词，表示能……的，有时会有一些变形。

vary——变化；variable——能变化的

adjust——调整；adjustable——可调节的

charge——充电；chargeable——能充电的

在单词前加上前缀，使单词有不同的含义和词性：

例如：charge——充电；recharge——重新充电；discharge——放电

large——大的；enlarge——使增大

act——作用，行动；react——反作用，反抗；interact——相互作用

单词加上后缀，改变其词性。

例如：help——帮助（动词）；helpful——有帮助的（形容词）

charge——充电（动词）；charger——充电器（名词）

react——反作用，反抗（动词）；reactance——反作用，反抗（名词）

在阅读英语时注意运用自己掌握的单词进行扩展记忆。

3. 词义引申

在阅读理解中，有时直译并不符合中文的习惯，这时要用词义引申，添加或删减，在保持原文意思不变的情况下，能更准确地表达含义且符合中文习惯。

The effect of resistance is constant regardless of frequency.

电阻值是与频率无关的常数。（这里 effect 可以略去。词汇删减）

Power plugs are male electrical connectors that fit into female electrical sockets.

电源插头可以插入电源插座。（词汇删减）

注意，在本教材的参考译文中，为了帮助理解，有时译得比较烦琐，例如这句话译作：

电源插头是电气接触头（俗称公插头），可以插入电源插座（俗称母插座）。一般翻译时，则能正确译出原意就可以了。

No band implies a tolerance of ± 20%, a silver band means the resistor has a tolerance of ± 10% and a gold band has the closest tolerance of ± 5%.

如果没有第四条，则电阻的精度为± 20%，如果第四条为银色，表示电阻的精度为± 10%，而金色表示精度为± 5%。（词汇增补）

从上文知，是在讨论第四条色带的作用，所以加上后说明比较清楚。另外英语中用了 implies, means 和 has 这三个不同的动词使叙述比较生动，但译成中文时都用"表示"比较清楚。

有时中英文在描述一些现象时用的方法是很不相同的，在专业文献中常常会遇到直译表述很不明确的问题，这时可结合我们的专业知识把原文的意思表达出来，把原来词的词性进行转换，有时更符合中文的习惯。

The charge current depends upon the technology and capacity of the battery being charged.

根据充电技术和电池容量的不同，充电电流的大小也不同。

2.4　Exercises

1. Put the Phrases into English（将下列词组译成英语）
(1) 理想电压源
(2) 比例常数
(3) 欧姆定律
(4) 电子器件
(5) 可变电容
(6) 阻挡直流
(7) 感抗
(8) 电磁能
(9) 交流电频率
(10) 输入阻抗

2. Put the Phrases into Chinese（将下列词组译成中文）
(1) linear resistor
(2) be inversely proportional to
(3) circuit diagram
(4) simplified capacitor
(5) pass ac current
(6) capacitive reactance
(7) as long as
(8) iron core
(9) variable inductors

(10) take into account

3. Sentence Translation (将下列句子译成中文)

(1) Suppose that some material is connected to the terminals of an ideal voltage source.

(2) Resistors are used as voltage dividers to reduce voltage for other device.

(3) There are many different types of capacitor but they all work in essentially the same way.

(4) The magnitude of the net charge Q on one plate is proportional to the applied voltage V.

(5) Impedance is more complex than resistance because the effects of capacitance and inductance vary with the frequency of the current passing through the circuit.

4. Translation (翻译)

Standard-type resistors usually maintain their value regardless of external conditions, such as voltage, temperature, and light. These types of resistors are referred to as linear resistors. There are other types of resistors referred to as nonlinear, whose resistance varies with temperature (thermistor 热敏电阻器), voltage (varistor 压敏电阻器) and light.

2.5 课文参考译文

基本元件通常是电阻器、电容器、电感器,它们并不总像看上去那样简单。即使在做一个最简单的项目(电路设计)之前我们都需要了解一些关于基本元件的知识。

2.5.1 电阻

第一个也是最常见的电子元件是电阻,几乎没有工作电路不使用它们。

设在某一器件的两端加上一个理想电压源,如图 2.1(a)所示,如果材料中流过的电流 $i(t)$ 总是正比于电压 $v(t)$,则这器件称为线性电阻,或简称为电阻。

因为对线性电阻来说,电压和电流总是成正比,其比值是常数,称为电阻(值)R。

流过一个电阻的电流与电阻两端的电压成正比,与电阻值成反比,这就是欧姆定律,用公式表示为:$v(t)=Ri(t)$。

电阻的单位(每安培伏特)称为欧姆,用希腊大写字母 Ω 表示。电阻的伏安特性如图2.1(b)所示。

电阻可用于限制流过器件的电流,保护器件使它不会被烧掉(并联分流);可用于(串联)分压,减小其他器件的电压,还可用作晶体管的偏置电路和作为电路的负载。图 2.2 为电路图中用到的各种电阻符号,矩形框的电阻符号欧洲用得比较多,折线式的电阻符号在日本和美国用得比较多。

2.5.2 电容

电容是可暂时存储电能的电子器件,几乎在所有复杂的电路中都可以找到电容。有很多各式各样的电容,但它们的基本工作原理是一样的。简单一点看电容是一对中间用称为电解质的绝缘材料分隔开的金属平板,这种一对平行板的简化电容结构也被选作为电容的电路符号如图 2.3(a)所示。有些电容的电容值是可以调节的,因此它们是可调电容[Fig 2.3(b)]。

如果在电容的两端(两层导体)加上电压,则电荷流入和聚集在一块极板上,同时相反极性的电荷在另一块极板上聚集,电流从另一端流出。

一块极板上聚集电荷 Q 的多少取决于所加的电压 V。因此有:$Q=CV$,其中 C 是电容值。

电容会阻挡直流通过,但却可以通过充电和放电让交流电流通过。电容对交流电有阻抗作用,称为容抗,容抗与电容值和所加交流电的频率有关,求容抗的公式为:$X_C=1/(2\pi f_C C)$,容抗的单位是欧姆 Ω。

2.5.3 电感

电感是一个电气元件,只要有电流流过电感,它就会以磁场的形式(临时)存储电磁能。线圈是最常用的电感器,但实际上,即使是一个直的金属丝也具有电感。给定长度的导线绕成线圈可以集中磁场并大大增加电感量。电感线圈中可以是空气,也可以有一个铁芯以增加电感量。电感的电路符号如图 2.4(a)所示。有些电感的电感量是可以调节的,在它们的线圈中有一个强磁的圆柱形铁芯,可以调节此铁芯在线圈中的位置(以改变电感量),称为可变电感。

电感总是反抗电流的变化,电感对直流电没有阻抗作用,但对交流电则有一个交流阻抗,称为感抗。感抗由电感值和交流电频率决定,求感抗的公式为 $X_L=2\pi f_L L$,感抗的单位是欧姆 Ω。

2.5.4 阻抗

(复)阻抗(符号为 Z)是反映电路中对电流的总的抵抗作用,换言之,是描述电路对电流的阻抗作用。它有点像电阻,但阻抗要考虑电容和电感的作用。阻抗的单位是欧姆,符号为 Ω。

阻抗比电阻复杂,因为电容和电感(对电流的阻碍)作用是会随着流过电路的电流的频率变化而变化的,这就意味着阻抗也会随频率变化。电阻值是与频率无关的常数。

阻抗这个词常用在没有电容和电感的简单电路中,(不是太正确)。例如谈到电路的输入阻抗或输出阻抗时,这一点容易把正在学电子技术的人搞糊涂。如果认为简单电路中阻抗是电阻的另一种说法,(就比较容易理解了)。

阻抗的值由四个电参量决定:电阻 R、电容 C、电感 L 和频率 f。

阻抗可以分成两部分:

电阻 R(这一部分与频率无关)。

电抗 X(这一部分与电容,电感有关,会随着频率变化)。

2.6 阅读材料参考译文

2.6.1 彩色条码电阻

低功率($\leq 2W$)器件常用彩色条码编号来区分它们的值和精度,一般有四色条码和五色条码两种，四色条码最常用，其精度为5%和10%，五色条码精度为1%甚至更高。

电阻的值可以根据色条码的颜色求出（见图2.5），颜色对应的值如表2.1所示。第1条是十位数，第2条是个位数，第3条是以10为基的指数值，第4条表示电阻的精度。如果没有第4条，则电阻的精度为±20%，如果第4条为银色，表示电阻的精度为±10%，而金色表示精度为±5%。

对有5色条码的电阻来说，第1条是百位数，第2条是十位数，第3条是个位数，第4条是以10为基的指数值。褐色、红色、绿色、蓝色和紫色只在5色条的电阻上用来表示精度。所有5色条的电阻都有表示精度的彩色条。

图2.6给出一些彩色条形码电阻，图2.6（a）中的褐—绿—灰—银—红条码电阻的阻值为1.58Ω，精度±2%。图2.6(b)中的黄—紫—橙—金条码电阻的阻值为47kΩ，精度±5%。

2.6.2　民用电源插头和插座

很多国家，民用电源是单相电源，用一根火线（相线）和一根中线将交流电送入民居。民用电源插头和插座是用来连接家用电器和家中所用的可移动的照明设施的。使家电和照明设施可以有电源供电，有电流流过。电源插头和插座有不同额定电压和电流额定值，其形状、尺寸和连接器的类型也不相同。在每个国家使用的插头和插座是按该国家标准设计的（图2.7）。许多插头和插座还有第三个接触头（孔）用来保护接地的。保护接地是当所接的电器设备出现（漏电）故障时引导电流流入大地的。

电源插头是电气接触头（俗称公插头）可以插入电源插座（俗称母插座）。圆形或扁平形的插头插入插孔或插槽中，使它们相互接触。插头通常有相线（或火线）、中线和接地的插头。很多插头并不区分相线和中线的插头。插头可以是用钢、青铜、锌、锡或镍材料制成的。

插座是另一种电气接触器，插座上有插槽或插孔让电源插头上的圆形或扁平形的插头插入，把电源中的电能输出给插头。插座通常设计成不允许电气标准不同的插头插入。有些插座有一组或多组插孔。

2.6.3　电池充电器

电池充电器是通过强迫输入电流给二次电池或（可充电）电池充电的设备。

充电电流的大小根据充电技术和电池的容量的不同而不同。例如，给汽车的12V电池充电的电流显然与给手机电池充电的电流大小是不一样的。

简单的充电器是把直流电源连接到待充电的电池上，这种简单的充电器并不会根据时间或电池电量的变化而改变它的输出。简单意味着比较便宜，但质量就需要权衡（即质量可能也一般）。简单充电器为了防止充电过度，一般充电时间比较长，即使这样，电池如果长时间放在简单充电器上会使电池因过度充电使其容量降低或坏掉。

连续充电器（慢充）就是一种以自放电速率进行慢充电的简单充电器。慢充是最慢的电池充电器，电池可以留在连续充电器中保持"充满"但不会过度充电。

定时充电器经过一段预定的时间后就停止充电。在20世纪90年代后期定时充电器是大容量的镍镉电池最常用的充电器。一般定时充电器和一组电池是捆绑销售的，如果较低容量电池用它充电则会过度充电。如果更高容量电池用它充电则充不满。

智能充电器的输出电流取决于电池的状态，一个智能充电器可以监测充电时电池的电压、温度和时间，输出当时最适当的充电电流。当电压、温度和时间的组合显示电池已充满时停止充电。

Unit 3 Semiconductor

3.1 Text

Semiconductors have had a monumental impact on our society. You find semiconductors at the heart of microprocessor chips as well as transistors(Fig 3.1). Anything that's computerized or uses radio waves depends on semiconductors.

Fig 3.1 semiconductor devices

3.1.1 Diode

A diode is the simplest possible semiconductor device, and is therefore an excellent beginning point if you want to understand how semiconductors work. A diode is an electrical device allowing current to move through it in one direction with far greater ease than in the other. Let's take a closer look at the simple battery-diode-lamp circuit (Fig 3.2). When placed in a simple battery-lamp circuit, the diode will either allow or prevent current through the lamp, depending on the polarity of the applied voltage.

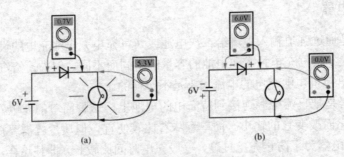

Fig 3.2 diode operation
(a) forward biased; (b) reversed biased

When the polarity of the battery is such that electrons are allowed to flow through the diode, the diode is said to be forward-biased [Fig 3.2(a)]. A forward-biased diode conducts current and drops a small voltage across it, leaving most of the battery voltage dropped across the lamp. Conversely, when the battery is "backward" and the diode blocks current, the diode is said to be reverse-biased [Fig 3.2(b)]. A reverse-biased diode drops all of the battery's voltage leaving none for the lamp. A diode may be thought of as like a switch: "closed" when forward-biased and "open" when reverse-biased.

The schematic symbol of the diode is shown in Fig 3.3(b) such that the anode (pointing end) corresponds to the P-type semiconductor at Fig 3.3 (a). The cathode bar, non-pointing end, corresponds to the N-type material at Fig 3.3 (a). Also note that the cathode stripe on the physical part Fig 3.3 (c) corresponds to the cathode on the symbol.

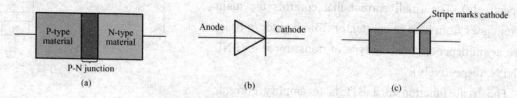

Fig 3.3 schematic symbol of the diode
(a) P-N junction representation; (b) Schematic symbol; (c) Real component appearance

The volt-ampere curve of a diode is shown in Fig 3.4, you should understand that the voltage dropped across a current-conducting diode does change with the amount of current going through it, but that this change is fairly small over a wide range of currents. This is why many textbooks simply say the voltage drop across a conducting, semiconductor diode remains constant at 0.7 volts for silicon and 0.3 volts for germanium.

A reverse-biased diode prevents current from going through it. In actuality, a very small amount of current can and does go through a reverse-biased diode, called the leakage current, but it can be ignored for most purposes. The ability of a diode to withstand reverse-bias voltages is

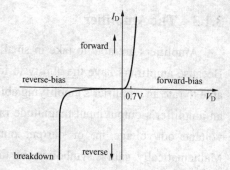

Fig 3.4 V-I curve of Diode

limited, as it is for any insulator. If the applied reverse-bias voltage becomes too great, the diode will experience a condition known as breakdown (Fig 3.4), which is usually destructive. A diode's maximum reverse-bias voltage rating is known as the Peak Inverse Voltage, or PIV.

3.1.2 BJT

The invention of the bipolar transistor in 1948 ushered in a revolution in electronics. Technical feats previously requiring relatively large, power-hungry vacuum tubes were suddenly achievable with tiny, power-thrifty specks of crystalline silicon. This revolution made possible the design and manufacture of lightweight, inexpensive electronic devices that we now take for granted. Understanding how transistors function is of paramount importance to anyone interested in modern electronics.

A bipolar Junction transistor (BJT) consists of a three-layer "sandwich" of doped (extrinsic) semiconductor materials, either P-N-P or N-P-N. An NPN BJT is shown in Fig 3.5; the figure is simplified compared with real devices. Two PN junctions are formed in the device: the collector-base junction and the emitter-base junction. We will see that the current flowing across one junction affects the current in the other junction. It is this interaction that BJT is known as a current controlled current device. In other words, they restrict the amount of current that can go through them according to a smaller, controlling current. The main current that is controlled goes from collector to emitter, or from emitter to collector, depending on the type of transistor it is (PNP or NPN, respectively). The small current that controls the main current goes from base to emitter, or from emitter to base, once again depending on the type of transistor it is (PNP or NPN, respectively).

The basic function of a BJT is to amplify current. This allows BJTs to be used as amplifiers or switches, giving them wide applicability in electronic equipment, including computers, televisions, mobile phones, audio amplifiers, industrial control, and radio transmitters.

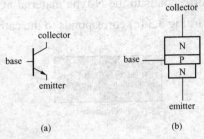

Fig 3.5   NPN BJT
(a) schematic symbol; (b) physical diagram

3.1.3   The Amplifier

Amplifiers are able to take in small-power signals and output signals of much greater power. Because amplifiers have the ability to increase the magnitude of an input signal, it is useful to be able to rate an amplifier's amplifying ability in terms of an output/input ratio. The technical term for an amplifier's output/input magnitude ratio is gain. As a ratio of equal units (power out/power in, voltage out/voltage in, or current out/current in), gain is naturally a unitless measurement. Mathematically, gain is symbolized by the capital letter "A".

For example, if an amplifier takes in an AC voltage signal measuring 0.2 volts RMS and outputs an AC voltage of 3 volts RMS, it has an AC voltage gain of 3 divided by 0.2, or 15:

$$A_v = \frac{V_{out}}{V_{in}} = \frac{3}{0.2} = 15$$

Correspondingly, if we know the gain of an amplifier and the magnitude of the input signal, we can calculate the magnitude of the output.

3.1.4   Operational Amplifiers

An operational amplifier (often op-amp or opamp) is a DC-coupled high-gain electronic voltage amplifier. The operational amplifier is a differential amplifier having both inverting input terminal v_1 and noninverting input terminal v_2 and one output v_o. The relationship between the output and the inputs is given by $v_o=A(v_2-v_1)$, where A is gain of the operational amplifier.

An ideal operational amplifier is modelled by the circuit shown in Fig 3.6(a), which contains a dependent voltage source. Note that since the input resistance $R=\infty$, when such an amplifier is connected to any circuit, no current will go into the input terminals(actually current is very small we ignore it). Also, since the output is the voltage across an ideal dependent voltage source, we have that regardless of what is connected to the output. For the sake of simplicity, the ideal amplifier having gain $A=\infty$ is often represented as shown in Fig 3.6(b).

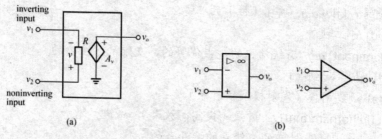

Fig 3.6   model and circuit symbol of an ideal OP amplifier
(a) model of OP amplifier; (b) circuit symbol

An ideal operational amplifier has the following characteristics:
(1) Infinite input impedance $R=\infty$.
(2) Infinite open-loop gain A for the differential signal $A=\infty$.
(3) Zero gain for the common-mode signal.
(4) Zero output impedance.

Review

(1) A forward-biased diode conducts current and a reverse-biased diode prevents current from going through it.
(2) BJT is known as a current controlled current device.
(3) Amplifier gain is an amplifier's output/input magnitude ratio.
(4) The operational amplifier is a differential amplifier having both inverting input terminal v_1 and noninverting input terminal v_2 and one output v_o.

Notes to the text

[1] bipolar-junction transistor (BJT)   双极性结式晶体管，中文常称晶体三极管。
[2] operational amplifier   运算放大器，简写作 OP Amplifier。

Technical Words

anode ['ænəud] n. [电]阳极，正极
biased ['baiəst] n. 偏置（电压） adj. 加偏置电压的，有偏见的
bipolar [bai'pəulə] adj. 有两极的，双极性的
cathode ['kæθəud] n. 负极，阴极
collector [kə'lektə] n. （计算机）集电极，收藏家，征收者
conversely ['kɔnvɜ:sli; kən'vɜ:sli] adv. 相反地

crystalline [ˈkrɪst(ə)laɪn] adj. 水晶的
destructive [dɪˈstrʌktɪv] adj. 破坏（性）的
diode [ˈdaɪəud] n. 二极管
dominant [ˈdɔmɪnənt] adj. 有统治权的，占优势的，支配的，主要的
doped [dopd] adj. 掺杂质的
emitter [iˈmitə] n. 发射体（发射极，辐射源）
enhancement [ɪnˈhɑːnsm(ə)nt] n. 增强（提高，放大），增进，增加
extrinsic [ekˈstrɪnsɪk] adj. 外在的，外表的，外来的
gain [geɪn] n. 增益，财物的增加，利润，收获 vt. 得到，增进，赚到 vi. 获利，增加增益
germanium [dʒɜːˈmeɪnɪəm] n. 锗
leakage [ˈliːkɪdʒ] n. 漏，泄漏，渗漏
microprocessor [maɪkrə(ʊ)ˈprəʊsesə] n. [计]微处理器
model [ˈmɔdl] n. 样式，模型 vt. 模仿 v. 模拟
monumental [mɒnjuˈment(ə)l] adj. 纪念碑的，纪念物的，不朽的，非常的
noninverting [ˈnɔnɪnˈvəːtɪŋ] n. 同相的，同相端的
paramount [ˈpærəmaʊnt] adj. 极为重要的
polarity [pə(ʊ)ˈlærɪtɪ] n. 极性
rate [reɪt] adj. 额定的
respectively [rɪˈspektɪvlɪ] adv. 分别地，各个地
semiconductor [ˌsemɪkənˈdʌktə] n. [物]半导体
silicon [ˈsɪlɪk(ə)n] n. 硅
usher [ˈʌʃə] n. 前驱 vt. 引导，展示，引座员
withstand [wɪðˈstænd] vt. 抵挡，经受住

Technical Phrases

forward biase	正向偏置
reversed biase	反向偏置
schematic symbol	图形符号
volt-ampere curve	伏安曲线
PN junction	PN 结
Peak Inverse Voltage	反向峰值电压
radio wave	无线电波
operational amplifier	集成运算放大器
noninverting input terminal	同相输入端
differential amplifier	差分放大器，差模放大器
common-mode signal	共模信号

3.2 Reading material

3.2.1 Application of the diode

Now we come to the most popular application of the diode: rectification. Simply defined, rectification is the conversion of alternating current (AC) to direct current (DC). This involves a device that only allows one-way flow of electrons. As we have seen, this is exactly what a semiconductor diode does. The simplest kind of rectifier circuit is the half-wave rectifier. It only allows one half of an AC waveform to pass through to the load (Fig 3.7).

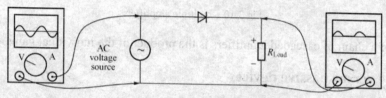

Fig 3.7 Half-wave rectifier circuit

For most power applications, half-wave rectification is insufficient for the task. The harmonic content of the rectifier's output waveform is very large and consequently difficult to filter. Furthermore, the AC power source only supplies power to the load once every half-cycle, meaning that much of its capacity is unused.

Another, more popular full-wave rectifier design exists, and it is built around a four-diode bridge configuration. For obvious reasons, this design is called a full-wave bridge (Fig 3.8).

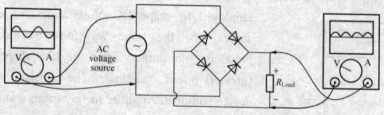

Fig 3.8 Full-wave bridge rectifier

Note that regardless of the polarity of the input, the current flows in the same direction through the load. That is, the negative half-cycle of source is a positive half-cycle at the load. The current flow is through two diodes in series for both polarities.

3.2.2 An example of amplifier

Fig 3.9 gives an amplifier formed by one transistor, resistors and capacitors, As an example, this one transistor amplifier is an "oldie but goodie". Note that by connecting the base-bias resistor

Rb to the collector you get two benefits: 1) the biasing cannot cause saturation or cut-off and 2) you introduce some negative feedback into the signal path, which reduces distortion. It's not as good as the Op-Amp circuit but it does work.

If multiple amplifiers are staged, their overall gain equal to the product (multiplication) of the individual gains (Fig 3.9). If a 1 V signal were applied to the input of the gain of 3 amplifier I (in Fig 3.10) a 3 V signal out of the first amplifier would be further amplified by a gain of 5 at the second stage yielding 15 V at the final output.

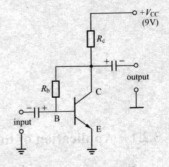

Fig 3.9 a voltage amplifier

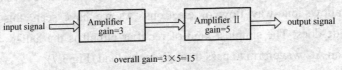

Fig 3.10 multiple amplifier

The gain of a chain of cascaded amplifiers is the product of the individual gains.

3.2.3 Active versus passive devices

An active device is any type of circuit component with the ability to electrically control electron flow (electricity controlling electricity). In order for a circuit to be properly called electronic, it must contain at least one active device. Components incapable of controlling current by means of another electrical signal are called passive devices. Resistors, capacitors, inductors, transformers, and even diodes are all considered passive devices. Active devices include, but are not limited to, vacuum tubes, transistors, MOS transistors and Operational Amplifiers.

While an amplifier can scale a small input signal to large output, its energy source is an external power supply (Fig 3.11). In other words, the current-controlling behavior of active devices is employed to shape DC power from the external power source into the same waveform as the input signal, producing an output signal of like shape but different (greater) power magnitude. The transistor or other active device within an amplifier merely forms a larger copy of the input signal waveform out of the DC power provided by a battery or other power source.

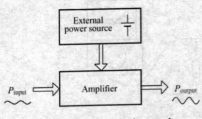

Fig 3.11 Active device need external power source

All active devices control the flow of electrons through them. Some active devices allow a voltage to control this current while other active devices allow another current to do the job. Devices utilizing a static voltage as the controlling signal are, not surprisingly, called voltage-controlled devices. Devices working on the principle of one current controlling another current are known as current-controlled devices. For the record, vacuum tubes are voltage-controlled devices while transistors are made as either voltage-controlled or current controlled types. The first type of transistor successfully demonstrated was a current-controlled device.

3.3 Knowledge about translation（非谓语动词 I）

在英语语法中，一句话中如果要叙述几个动作时先选其中主要的动作作为谓语，用动词表示，其余动作要用动词的变形表示，称为非谓语动词。非谓语动词通常有三种：V-ing、V-ed 和 to V。

V-ing 形式的动词在一些语法书中分为动名词与现在分词，但现在的趋势是不加区分，本书中不加区分，只要能准确理解句子中的意思没有必要去辨别它是动名词还是现在分词，V-ing 在句中可以作为表语、主语、宾语、定语，同时保留了动词性，因此可带有宾语和状语。

（1）V-ing 作定语：

单个 V-ing 作定语一般放在名词前面（也可以放在后面），V-ing 短语作定语一般放在名词之后。且 V-ing 本身含有主动、进行的意思，表示动作是由所修饰的名词主动发出的。

Devices **utilizing** a static voltage as the **controlling** signal are called voltage-controlled devices.

利用一个电压作为控制信号的器件叫电压控制器件。

这里 V-ing 短语作定语，utilizing（利用）的动作是由器件发出的。这里的 static 是静止的意思，可以译成一定的电压，但根据专业知识，不翻译出来也可以。

A diode is an electrical device **allowing** current to move through it in one direction with far greater ease than in the other.

二极管是这样一种器件，一个方向流过二极管的电流远大于另一个方向的。

这里 V-ing 短语作定语，allowing 由 device 发出的动作，根据专业知识，也可以译成：二极管具有电流单向导通的特点。

All **moving** bodies have energy.

所有运动的物体都有能量。

这里 V-ing 作定语，放在所定义名词的前面。

（2）作状语：

V-ing 短语作状语时，往往具有时间、条件、原因、结果、方式、补充说明等含义，它可放在句首、句中或句尾，通常它的逻辑主体就是句中的主语。

When placed in a simple battery-lamp circuit, the diode will either allow or prevent current through the lamp, **depending** on the polarity of the applied voltage.

当在一个简单的电池—电灯电路中串联一个二极管后，根据所加的电池电源的极性的不同，二极管会出现导通电流或阻碍电流流过电灯的现象。

这里 V-ing 作状语，说明条件。

Being negative, electrons move always from negative to positive.

电子是负的，所以总是从负（极）向正（极）运动。（V-ing 短语作原因状语）

分词短语作状语时，前面可用 when、while、if、unless、though 等连词来加强时间、条件等含义。

When measuring current, the circuit must be opened and the meter inserted in series with the circuit or component to be measured.

当测量电流时，必须断开电路，将万用表与待测电路或元器件串联。

（3）作主语或宾语：

Heating the water changes it into vapor.

把水加热可以使水变为蒸汽。（V-ing 短语作主语）

It may also have a polarity switch to facilitate **reversing** the test leads.

还有一个极性开关可以很方便交换测试笔的极性。（V-ing 短语作 facilitate 的宾语）

（4）作主语或宾语的补足语：

We put a hand above an electric fire and feel the hot air **rising**.

我们把手放在电炉的上方，就会感觉到热空气在上升。

（5）与 with (without) 连用：

在科技文章中，常用 with (without) + 名词 + 分词 结构用作补充说明。这种结构中，with 没有词汇的意思，表示一种伴随情况，可根据具体情况进行理解。

The density of air varies directly as pressure, with temperature **being** constant.

在温度不变时，空气密度与压力成正比。

3.4 Exercises

1. Put the Phrases into English（将下列词组译成英语）

(1) P 型半导体
(2) 微处理器芯片
(3) 反向偏置
(4) 反相输入
(5) 集电极
(6) 可控电压源
(7) 理想放大器
(8) 输出阻抗
(9) 发射极
(10) 三极管

2. Put the Phrases into Chinese（将下列词组译成中文）

(1) the polarity of the battery
(2) across the lamp

(3) schematic symbol

(4) cathode bar

(5) a three-layer "sandwich"

(6) operational amplifier (OP-amplifier)

(7) PN junction

(8) collector-base junction

(9) semiconductor diode

(10) modern electronics

3. Sentence Translation (将下列句子译成中文)

(1) When placed in a simple battery-lamp circuit, the diode will either allow or prevent current through the lamp, depending on the polarity of the applied voltage.

(2) A diode may be thought of as like a switch: "closed" when forward-biased and "open" when reverse-biased.

(3) Understanding how transistors function is of importance to an electronics engineer.

(4) The small current that controls the main current goes from base to emitter in a NPN transistor.

(5) The operational amplifier is a differential amplifier having both inverting input terminal v_1 and noninverting input terminal v_2 and one output v_o.

4. Translation (翻译)

Bipolar transistors are so named because the controlled current must go through two types of semiconductor material: P and N. Transistors function as current regulators by allowing a small current to control a larger current. The amount of current allowed between collector and emitter is primarily determined by the amount of current moving between base and emitter.

3.5　课文参考译文

半导体技术对我们的社会（发展）有着非常大的影响，微处理器芯片中用到半导体，晶体管也用到半导体（图3.1），任何用到计算机或无线电波的技术都要用到半导体。

3.5.1　二极管

二极管是最简单的半导体器件，如果想要了解半导体是如何工作的，最好是从二极管开始。一个方向流过二极管的电流远大于另一个方向的（即二极管具有电流单向导通的特点）。先观察一个简单的电池—二极管—电灯的电路（图3.2）。当在一个简单的电池—电灯电路中串联一个二极管后，根据所加的电池电源极性的不同，二极管会出现导通电流或阻碍电流流过电灯的现象。

当所加电池的极性使得电子能流过二极管，就称为正向偏置［图3.2（a）］。正向偏置的二极管导通（电流）并有一个很小的正向压降，而让绝大部分电压加在电灯上。反之，当电池"反接"二极管阻挡电流，则称二极管反向偏置。反向偏置二极管两端的电压降等于电池的全部电压，电灯两端没有电压。可以把二极管看成一个开关：正向偏置时"合上"，反向偏置时"断开"。

二极管的电路符号如图 3.3（b）所示，与 P 型半导体相接的一端是阳极或正极（箭头端），与 N 型材料相应的一端是阴极（一竖条，非箭头端）。另外在实际的二极管外壳上有一条线表示这端为阴极，如图 3.3（c）所示。

二极管的伏—安（特性）曲线如图 3.4 所示，要知道一个导通的二极管其两端的电压是随着流过二极管的电流的不同而不同的，但在电流变化范围内这个电压变化相当小，因此很多教材都简单地说一个导通的二极管两端的电压是一个常数，这常数对硅管来说是 0.7V，对锗管来说是 0.3V。

二极管反向偏置时阻挡电流流过。实际上是有一个很小的电流流过二极管的，称为漏电流，但对大部分应用来说这个漏电流是可以忽略不计的。二极管承受的反向偏置电压也是有限的，就如任何绝缘材料承受的电压都是有限的一样。如果所加的反向偏置电压太大了，二极管就会击穿，通常二极管击穿后就坏了。二极管所能承受的最大反向偏置电压称为反向峰值电压，记为 PIV。

3.5.2 三极管

1948 年发明的双极性晶体管（又称三极管）是电子技术方面的一场革命，以前所用的体积相当大且耗电的真空管突然被微小的、省电的晶体管取代了。这场革命使得设计和制造今天所用的轻巧且便宜的电子设备成为可能，对想要了解现代电子学的人来说，了解晶体管的性能是十分重要的。

双极性晶体管（我国常称三极管）由三层掺杂的半导体材料 NPN 或 PNP（类似于一个"三明治"）组成，图 3.5 为一个 NPN 型的三极管，是实际器件的简化图。器件中形成了两个 PN 结：集电极—基极结和发射极—基极结。可以观察到流过一个结的电流影响流过另一个结的电流，就是这种相互作用三极管被称为是一个电流控制电流的器件。换言之，一个较小的控制电流限制（控制）了可以通过三极管的电流的大小。从集电极到发射极的电流（NPN 型）或从发射极到集电极的电流（PNP 型）是受控的大电流，大电流的方向由晶体管的类型决定，而控制这个电流的小电流是从基极流向发射极（NPN 型）或从发射极流向基极（PNP 型），同样小电流的方向也由晶体管的类型决定。

三极管的基本功能是放大电流。这个功能使得三极管可以构成放大器或开关电路，因此三极管在包括计算机、电视、手机、音频放大器、工业控制和无线电发射机等电子设备中应用非常广泛。

3.5.3 放大器

放大器能接收小功率信号，并输出比较大的功率信号。因为放大器有增加输入信号幅度的能力，因此可以用放大器输出信号/输入信号的比值作为衡量一个放大器的放大能力的指标。一个放大器输出信号/输入信号的比值在专业词汇中称为增益（放大倍数）。作为同单位的比值（功率输出/功率输入，电压输出/电压输入，或电流输出/电流输入），增益是一个没有单位的量。数学上,增益用的大写字母"A"表示。

例如:如果一个放大器输入一个有效值为 0.2V 的交流电压信号,输出有效值为 3V 的交流电压信号,则它的交流电压增益(放大倍数)为 3 除以 0.2 或 15。

$$A_v = \frac{V_{\text{out}}}{V_{\text{in}}} = \frac{3}{0.2} = 15$$

相应地,如果我们知道一个放大器的增益和输入信号的幅度,可以求出输出信号的幅度。

3.5.4 运算放大器

运算放大器(通常简写成 op-amp 或 opamp)是一种直流耦合的高增益的电子电压放大器。运算放大器是一个差分放大器,它有一个反相输入端 v_1、一个同相输入端 v_2 和一个输出端 v_o。输出端和输入端的关系可以为:$v_o = A(v_2 - v_1)$,其中 A 为运算放大器的电压放大倍数。

一个理想的运算放大器模型如图 3.6(a)所示,其中含有一个受控电压源。因为输入电阻 $R = \infty$,因此运算放大器接在任何电路上输入端都没有电流流入(实际为很小,可以忽略不计)。又因为输出端与一个理想的受控电压源相连,因此输出端可以接任何电路。简化后,理想运算放大器的增益(电压放大倍数)$A = \infty$,常用图 3.6(b)所示方法表示。

一个理想运算放大器有以下特点:
(1)输入阻抗无限大,$R = \infty$。
(2)开环差动放大倍数无限大,$A = \infty$。
(3)对共模信号的增益(共模放大倍数)为零。
(4)输出阻抗为零。

3.6 阅读材料参考译文

3.6.1 二极管的应用

现在讨论二极管最基本的应用:整流。简单地说,整流是把交流电(AC)转换成直流电(DC)。这就用到只允许单方向流过电流的器件,就如我们所知道的,这种器件就是半导体二极管。最简单的整流电路是单波整流器,它只允许一半交流电流过负载(图 3.7)。

在很多电源应用中,半波整流是不够的,整流器的输出波形中的谐波(非直流)分量很大,很难滤去,再加上交流电源只有半周对负载供电,(其能力)没有得到充分的利用。

另一种更常用的设计是全波整流电路,它由四个二极管构成,称为全波桥式整流器(图 3.8)。

可以看到无论输入电压是正极性还是负极性,流过负载的电流是同样方向的,即在电源负半周时流过负载的电流方向与电源正半周时流过负载的电流方向是相同的。电源为不同极性时都分别有两个二极管与负载相串联使电流导通。

3.6.2 放大器举例

图 3.9 给出一个用三极管、电阻和电容构成的放大器（放大电路）。作为一个例子，这款三极管放大器虽然比较老性能却比较好，图中把偏置电阻 R_b 直接连接到集电极的接法，有两个好处，一是这种偏置不可能引起三极管饱和或截止，二是这个电路构成的负反馈可以减小电路中的干扰。这个电路可能不如集成运放电路（性能好），但它确实是一个实用的电路。

如果是多级放大器级连，则总的增益等于各个放大器各自增益的乘积（图 3.10）。如果把一个 1V 的信号加到一个增益为 3 的放大器 I 的输入端，则第一个放大器的输出信号 3V 将被一个增益为 5 的第二级放大器再次放大，最终输出一个 15V 的信号。

一个（多级）级联放大器的增益是各个放大器增益的积。

3.6.3 有源器件与无源器件

有源器件是一种具有电量控制电子流（电量控制电量）的电路器件。如果一个电路称为电子电路，那么它至少含有一个有源器件。不能用另一种电信号控制电流的器件称为无源器件。电阻、电容、电感、变压器甚至二极管都可以看成是无源器件。有源器件包括（但不限于）真空管、晶体管、MOS 晶体管（场效应管）和运算放大器。

当一个放大器可以把一个小的输入信号按比例放大输出，它的能源是要另外提供的（图 3.11）。换言之，有源器件的电流控制性能是把来自外接电源的直流能量转换成与输入信号相同的波形，产生一个形状相同，但不同（较大）功率的输出信号。晶体管或其他放大器内部的有源器件只是利用电池或其他电源提供的直流电能形成输入信号波形的一个比较大的复制品（输出）。

有源器件都能控制流过器件的电子流，有些有源器件允许用电压控制电流而另一些有源器件则允许另一个（小）电流控制电流。用一个电压作为控制信号的器件，毫不奇怪地，叫电压控制器件。而用一个电流控制另一个电流的器件称为电流控制器件。准确地说，真空管是电压控制器件，而晶体管有电压控制型（MOS 管或称场效应晶体管）或电流控制型（三极管）的。最早制成的晶体管是电流控制器件。

Unit 4 Analysis of an electric circuit

4.1 Text

The study of electric circuits is fundamental in electrical engineering education and can be quite valuable in other disciplines as well. The skills acquired not only are useful in such electrical engineering areas as electronics, communications, microwaves, control, and power systems but also can be employed in other seemingly different fields.

By an electric circuit or network we mean a collection of electrical devices (for example, voltage and current sources, resistors, inductors, capacitors, transformers, amplifiers, and transistors) that are interconnected in some manner. The process by which we determine a variable (either voltage or current) of a circuit is called analysis.

4.1.1 Kirchhoff's Current Law

It is a consequence of the work of the German physicist Gustav Kirchhoff[1] (1824—1887) that enables us to analyze any complex circuit.

For a given circuit, a connection of two or more elements shall be called a node (Fig 4.1). We now present the first of Kirchhoff's two laws, his current law (KCL), which is essentially the law of conservation of electric charge:

At any node of a circuit, at every instant of time, the sum of the currents into the node is equal to the sum of the currents out of the node.

Fig 4.1 node of a circuit

$$\sum i_{in} = \sum i_{out}$$

For example: at the node in Fig 4.1, we have: $i_1+i_4=i_2+i_3$.

An alternative, but equivalent, form of KCL can be obtained by considering currents directed into a node to be positive in sense and currents directed out of a node to be negative in sense, under this circumstance, the alternative form of KCL can be stated as follows:

$$\sum i = 0$$

At any node of a circuit, at every instant of time, the currents algebraically sum to zero.

For example: at the node in Fig 4.1, we have: $-i_1+i_2+i_3-i_4=0$.

4.1.2 Kirchhoff's Voltage Law (KVL)

We now present the second of Kirchhoff's laws-the voltage law. To do this, we must introduce the concept of a "loop". Starting at any node n in a circuit, we form a loop by traversing through elements and returning to the starting node n, and never encountering any other node more than once (Fig 4.2).Kirchhoff's voltage law (KVL) is:

In traversing any loop in any circuit, at every instant of time, the sum of the voltages having one polarity equals the sum of the voltages having the opposite polarity.

$$\sum v_+ = \sum v_-$$

For example: in the loop shown in fig 4.2, we have: $v_2+v_4=v_1+v_3$.

An alternative statement of KVL can be obtained by considering voltages across elements that are traversed from plus to minus to be positive in sense and voltages across elements that are traversed from minus to plus to be negative in sense (or vice versa). Under this circumstance, KVL has the following alternative form.

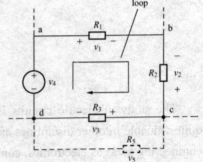

Fig 4.2 a loop of the circuit

$$\sum v = 0$$

Around any loop in a circuit, at every instant of time, the voltages algebraically sum to zero.

For example: shown in fig 4.2, we have: $v_1-v_2+v_3-v_4=0$.

4.1.3 Nodal analysis

In analysis of electrical circuits, there are several distinct approaches that we can take. In the one we write a set of simultaneous equations in which the variables are voltage; this is known as nodal analysis.

Given a circuit with n nodes and no voltage sources, proceed as follows:

(1) Select any node as the reference node.

(2) Label the remaining $n-1$ nodes (e.g., $v_1, v_2, \cdots, v_{n-1}$).

(3) Apply KCL at each nonreference node by summing the currents out of the nodes, Use Ohm's law to express the currents through resistors in terms of the node voltages.

(4) Solve the resulting set of $n-1$ simultaneous equations for the node voltages.

Take the circuit in Fig 4.3 as an example:

(1) select a reference point

(2) label A,B point, V_A, V_B.

(3) by KCL, we have:

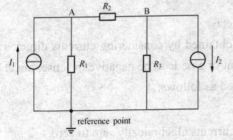

Fig 4.3 example circuit

$$\frac{V_A}{R_1} + \frac{V_A - V_B}{R_2} - I_1 = 0$$

$$\frac{V_B}{R_3} + \frac{V_B - V_A}{R_2} + I_2 = 0$$

(4) solve the equations.

Review

(1) The process by which we determine a variable (either voltage or current) of a circuit is called analysis.
(2) At any node of a circuit, at every instant of time, the sum of the currents into the node is equal to the sum of the currents out of the node.
(3) In traversing any loop in any circuit, at every instant of time, the sum of the voltages having one polarity equals the sum of the voltages having the opposite polarity.

Notes to the Text

[1] Gustav Kirchhoff 人名，德国物理学家（1824—1887）他给出了电路分析的两个基本定律，分别称为基尔霍夫电流定律和基尔霍夫电压定律。

Technical Word

algebraical [ˌældʒɪˈbreɪɪk,-kəl] a. 代数的（等于 algebraic）
alternative [ɔːlˈtɜːnətɪv] adj. 交替的，轮流的，预备的 v. 交替，轮流，改变
analysis [əˈnælɪsɪs] n. 分析，分解
circumstance [ˈsɜːkəmstəns] n. 环境，详情，境况
conservation [ˌkɒnsəˈveɪʃ(ə)n] n. 保存，保持，守恒
discipline [ˈdɪsɪplɪn] n. 学科
encounter [ɪnˈkaʊntə] v. 遇到，相遇   n. 遭遇，遭遇战
equivalent [ɪˈkwɪv(ə)l(ə)nt] adj. 相等的，相当的   n. 等价物，相等物
fundamental [fʌndəˈment(ə)l] adj. 基础的，基本的   n. 基本原则，基本原理
instant [ˈɪnst(ə)nt] adj. 立即的，直接的，紧迫的，刻不容缓的，（食品）速溶的，方便的，即时的
loop [luːp] 回路
microwave [ˈmaɪkrə(ʊ)weɪv] n. 微波（波长为 1 毫米至 30 厘米的高频电磁波）
negative [ˈnegətɪv] n. 否定，负数，底片   adj. 否定的，消极的，负的，阴性的   vt. 否定，拒绝（接受）
network [ˈnetwɜːk] n. 网络，网状物，电路
node [nəʊd] 节点

nonreference [nɒnˈref(ə)r(ə)ns] 非参考点
present [ˈprez(ə)nt] vt. 介绍，提出，呈现
simultaneous [ˌsim(ə)lˈteiniəs] adj. 同时的
traverse [ˈtrævəs] vt. 横过，经过，在……来回移动　vi. 横越，来回移动
variable [ˈvɛriəbl] n. 变量　adj. 可随意调节的；可变的，（数）变量的

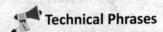

Technical Phrases

conservation of electric charge	电荷守恒
nodal analysis	节点电位（分析）法
from plus to minus	从正到负
in sense	在……意义上，假设

4.2 Reading material

4.2.1 Mesh analysis

In another way to analyze the circuit we write a set of simultaneous equations in which the variables are currents, this is known as mesh analysis.

Given a planar circuit with m meshes and no current sources, proceed as follows:

(1) Place clockwise mesh currents (e.g., $i_1, i_2, \cdots, i_m$) in the m (finite) meshes.

(2) Apply KVL to each of the m meshes by traversing each mesh in the clockwise direction. Use Ohm's law to express the voltages across resistors in terms of the mesh currents.

(3) Solve the resulting set of m simultaneous equations for the mesh currents.

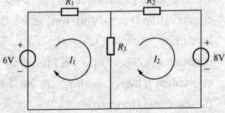

Fig 4.4　example circuit

Take the circuit shown in Fig 4.4 as an example:

(1) place mesh currents I_1, I_2.

(2) by KVL, for the first loop, we have:

$$I_1 R_1 + (I_1 - I_2) R_3 - 6 = 0$$

Second loop:

$$I_2 R_2 + 8 + (I_2 - I_1) R_3 = 0$$

(3) solve the equations.

4.2.2 Analog signal

An analog or analogue signal is any time continuous signal where some time varying feature of the signal is a representation of some other time varying quantity. An analog signal uses some property of the medium to convey the signal's information. Electrically, the property most commonly used is voltage followed closely by frequency, current, and charge.

Any information may be conveyed by an analog signal; often such a signal is a measured response to changes in physical phenomena, such as sound, light, temperature, position, or pressure, and is achieved using a sensor.

For example, in sound recording, fluctuations in air pressure (that is to say, sound) strike the diaphragm of a microphone which causes corresponding fluctuations in a voltage or the current in an electric circuit. The voltage or the current is said to be an "analog" of the sound.

Any measured analog signal must theoretically have noise and a finite slew rate. Therefore, both analog and digital systems are subject to limitations in resolution and bandwidth. In practice, as analog systems become more complex, effects such as non-linearity and noise ultimately degrade analog resolution. In analog systems, it is difficult to detect when such degradation occurs. However, in digital systems, degradation can not only be detected but corrected as well.

4.2.3 Thévenin's theorem[1]

Suppose that a load resistor R_L is connected to an arbitrary (in the sense that it contains only linear elements) circuit as shown in Fig 4.5(a). What value of the load R_L will absorb the maximum amount of power? Knowing the particular circuit, we can use nodal or mesh analysis to obtain an expression for the power absorbed by R_L, then take the derivative of this expression to determine what value of R_L results in maximum power. The effort required for such an approach can be quite great. Fortunately, though, a remarkable and important circuit theory concept states that as far as R_L is concerned, the arbitrary circuit shown in Fig4.5 (a) behaves as though it is a single independent voltage source in series with a single resistance [Fig 4.5 (c)].

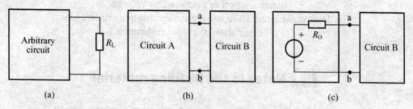

Fig 4.5　Thévenin's theorem
(a) arbitrary circuit and its associated load; (b) circuit partitioned into two parts;
(c) application of Thévenin's

Once we determine the values of this source and this resistance, we simply apply the results on maximum power transfer.

Suppose we are given an arbitrary circuit containing any or all of the following elements resistors, voltage sources, current sources. (The sources can be dependent as well as independent) Let us identify a pair of nodes, say node a and node b, so that the circuit can be partitioned into two pars as shown in Fig4.5 (b), Furthermore suppose that circuit A contains no dependent source that is dependent on a variable in circuit B, and vice versa. Then we can replace circuit A by an appropriate independent voltage source, call it u_{OC}, in series with an appropriate resistance, call it R_O, and the effect on circuit B is the same as that produced by circuit A. This voltage source and resistance series combination is called the Thévenin equivalent of the circuit A. In other words, circuit A in Fig4.5 (a) and the circuit in the left box in Fig4.5 (c) have the same effect on the circuit B. This result is known as Thévenin's theorem, and is one of the more useful and significant concepts in circuit theory.

To obtain the voltage u_{OC}–called the open-circuit voltage-remove circuit B from circuit A, and determine the voltage between nodes a and b. This voltage, as shown in Fig4.6(a) is u_{OC}.

To obtain the resistance R_O called the Thévening-equivalent resistance or the output resistance of circuit A–again remove circuit B from circuit A. Next, set all independent sources in circuit A to zero, Leave the dependent sources as is! (A zero voltage source is equivalent a short circuit, and a zero current source is equivalent to an open circuit) Now determine the resistance between nodes a and b–this is R_O as shown in Fig4.6 (b).

If circuit A contains no dependent sources, when all independent sources are set to zero, the result may be simply a series-parallel resistive network in this case, however, R_O can be found by applying an independent source between nodes a and b and then by taking the ratio of voltage to current. This procedure is depicted in Fig4.6 (c), for the most part it doesn't matter whether u_O is applied and i_O is calculated or vice versa.

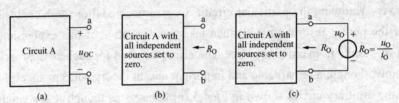

Fig 4.6　find the Thévenin equivalent circuit
(a) determination of open-circuit voltage;
(b) determination of Thévenin equivalent (output) resistance;
(c) determination of output resistance

📖 Notes to the Reading material

[1] Léon Charles Thévenin（1857—1926）—法国通信工程师，他提出了著名的戴维宁定理，有的教材译作戴维南定理。

4.3　Knowledge about translation（非谓语动词Ⅱ）

1. to V 形式
to V 又称动词不定式，兼有名词、形容词、副词特点，也保留了动词性，可用作主语、宾语、表语、定语、状语。常用来表示具体的（特别是未来的）一次性动作。

Resistor R_F is made variable to be able to adjust sufficient feedback voltage **to cause oscillation**.

电阻是可调的，可调节到有足够的反馈电压以引起**电路的**振荡。

这里 is made 不译，"电路的"为增补词语。用了 to V 表示一次性的动作，未来的动作。

It is a consequence of the work of the German physicist Gustav Kirchhoff (1824—1887) that enables us **to analyze** any complex circuit.

正是德国物理学家 Gustav Kirchhoff（1824—1887）研究的结论，使我们可以分析任何复杂的电路。

这里用了 It is…that，用来强调，To V 作宾语补足语），参考译文中译作：德国物理学家 Gustav Kirchhoff（1824—1887）研究得出了基尔霍夫定律，它可以用来分析任何复杂的电路。这样比较符合中文的习惯。

It is very difficult **to measure the passing current in insulators**.

测量绝缘体中通过的电流是很困难的。

不定式短语作主语时，尤其是不定式短语比较长时，往往引入 it 作形式主语，而把不定式短语放在谓语动词的后面。

To do this, we must introduce the concept of a "loop".

为了这么做，必须引入"回路"的概念。

这里不定式短语作状语，由上下文也可以译成：为了给出基尔霍夫第二定律，必须引入"回路"的概念。

Tesla's work on induction motors and polyphase systems influenced the field for years **to come**.

特斯拉关于感应电动机和多相输电系统的研究到今天还影响着电气工程领域。

这里 to V 作定语，定 years，未来的年代。

to V 与 V-ing 的区别：

在翻译中要特别注意 to V 与 V-ing 的区别，否则易造成曲解。

如：Stop to smoke. 停下来抽一支烟（一次性动作）

Stop smoking. 戒烟（终止这一经常性的动作）

I forgot to do it. 我忘记做这件事了（事没有做）

I forgot doing it. 我忘记做过这件事了（事已做了）

2. V-ed 形式

动词的 V-ed 形式与 be 结合构成被动态，与 have (had) 结合构成完成时，因此 V-ed 形式本身含有被动与完成的意思，它可在句子中担任定语、状语等，保留了动词性，表示这一动作是已完成的或所修饰名词所（被动）接受的。

Multidigit displays consist of two or more seven-segment displays **contained** in a single package or module.

多位数码显示器由两个或更多的七段显示器（被封装在一个单元中）组成。

（1）V-ed 作定语

单个 V-ed 作定语一般放在名词前面（也可以放在后面），V-ed 短语作定语一般放在名词之后。在作定语时，分词在意思上接近一个定语从句。

The skills acquired are useful in electrical engineering areas.

要求的技能在电气工程领域是十分有用的。（V-ed 作定语）

For a given circuit, a connection of two or more elements shall be called a node.

对一个给定的电路，两个或更多元件的连接点称为节点。（V-ed 作定语）

A semiconductor diode consists of a PN junction **made of semiconductor material**.

一个半导体二极管是由一个半导体材料制成的 PN 结构成的。

（2）V-ed 作状语

V-ed 作状语可表示动作发生的背景和情况。

Compared to a conventional process control system, number of wires needed for connections is reduced by 80%.

与传统的控制柜系统比较，所需的连接线减小了 80%。

Tonight, **lit** by countless electric lights, all the halls are as bright as day.

今晚无数个电灯把所有的大厅照耀得如同白昼。

3．V-ing 与 V-ed 的区别

正如前面已提到的，V-ing 通常有主动的意思，V-ed 通常有被动的意思，在阅读翻译时要注意这一点，才能区分动作的发出者和对象。试比较：

I heard someone opening the door. 我听见有人开门。

I heard the door opened. 我听见门被（谁）打开了。

The switched (controlled) current goes between emitter and collector, while the controlling current goes between emitter and base.

当控制电流流过发射极和基极时，被控电流流过发射极和集电极。

4.4　Exercises

1. Put the Phrases into English (将下列词组译成英语)

(1) 电路分析

(2) 电荷守恒定律

(3) 电路节点

(4) 在每一瞬时

(5) 电流代数和

(6) 形成一个回路

(7) 相反的极性

(8) 元件两端的电压

(9) 参考节点

(10) 求解方程

2. Put the Phrases into Chinese (将下列词组译成中文)

(1) in other disciplines

(2) determine a variable

(3) either voltage or current

(4) the sum of the currents into the node

(5) under this circumstance

(6) starting at any node n

(7) from minus to plus

(8) shown in fig 4.2

(9) in terms of the node voltages

(10) apply KCL

3. Sentence Translation (将下列句子译成中文)

(1) The study of electric circuits is fundamental in electrical engineering education and can be quite valuable in other disciplines as well.

(2) The process by which we determine a variable (either voltage or current) of a circuit is called analysis.

(3) For a given circuit, a connection of two or more elements shall be called a node.

(4) In analysis of electrical circuits, there are several distinct approaches that we can take.

(5) Apply KCL at each nonreference node by summing the currents out of the nodes.

4. Translation (翻译)

A multimeter or a multitester, also known as a volt/ohm meter or VOM, is an electronic measuring instrument that combines several functions in one unit. A standard multimeter may include features such as the ability to measure voltage, current and resistance. There are two categories of multimeters, analog multimeters (or analogue multimeters in British English) and digital multimeters (often abbreviated DMM or DVOM).

4.5 课文参考译文

电路的学习研究是电气工程教育的基础，在其他学科中也十分重要。在电路分析中所学到的技能不仅在这些电气工程领域如电子学、通信、微波、控制和电力系统中十分有用，而且也可以应用在其他不同的领域。

一个电路或网络指的是一组电气器件设备（如电压电流源、电阻、电感、电容、变压器、放大器和晶体管）以某种方式相互连接在一起构成的。求解电路中的变量（无论是电压还是电流）的过程称为电路分析。

4.5.1 基尔霍夫电流定律

德国物理学家 Gustav Kirchhoff（在 1824—1887）的研究得出了基尔霍夫定律，它可以用来分析任何复杂的电路。

对一个给定的电路，两个或更多元件的连接点称为节点（图 4.1）。现在先给出基尔霍夫两个定律中的第一个，基尔霍夫电流定律（KCL）。KCL 是基于电荷守恒的定律得到的。

在电路的任何一个节点，在任何一瞬时，流入节点电流的总和等于流出节点电流的总和。

$$\sum i_{in} = \sum i_{out}$$

例如图 4.1 所示的节点，有：$i_1+i_4=i_2+i_3$。

KCL 定律还有一种等价的说法，是通过假设流入结点的电流方向为正，流出节点的电流方向为负，在这种假设下，KCL 的另一个表达形式为：

$$\sum i = 0$$

电路的任一节点在任一瞬时，电流的代数和等于 0。

例如图 4.1 所示的节点，有：$-i_1+i_2+i_3-i_4=0$。

4.5.2 基尔霍夫电压定律（KVL）

为了给出基尔霍夫的第二定律，先引入"回路"的概念。从电路中的任一节点 n 出发，通过电路中每一个元件，每个其他节点只能通过一次，再回到电路的出发节点 n 形成一个闭合回路。则基尔霍夫电压定律是：

沿着电路的任一闭合回路，在任一瞬时，同极性电压（例如电位上升）之和等于另一反极性电压（例如电位下降）之和。

$$\sum v_+ = \sum v_-$$

例如图 4.2 所示的回路，有：$v_2+v_4=v_1+v_3$。

通过假定元件两端的电压是从正到负为正电压，元件两端的电压是从负到正为负电压，可以得到基尔霍夫电压定律另一种描述方法。这时，基尔霍夫电压定律也可以表述成下列形式：

$$\sum v = 0$$

绕电路中的任一闭合回路，在任一瞬时，电压的代数和等于 0。

例如图 4.1 所示的回路，有：$v_1-v_2+v_3-v_4=0$。

4.5.3 节点（电位）分析法

分析电路有几种方法，其中一种是列出同一时刻的方程组，其变量是电压，这种方法称为节点（电位）分析法。

对一个有 n 个节点，不含电压源的电路，节点（电位）分析法的步骤如下：

（1）选取任何一个节点作为参考点。

（2）对其他节点标出假设电位，如 $v_1, v_2, \cdots, v_{n-1}$。

（3）对每个非参考节点运用 KCL，即求出流出节点的电流和（应等于 0），根据节点电位用欧姆定律表示流过电阻的电流，（列出 $n-1$ 个方程）。

（4）解 $n-1$ 个方程，求出节点电压。

以图 4.3 中的电路为例：

（1）选参考点。

（2）标出 A、B 点，其电位为 V_A、V_B。
（3）由 KCL，得

$$\frac{V_A}{R_1} + \frac{V_A - V_B}{R_2} - I_1 = 0$$

$$\frac{V_B}{R_3} + \frac{V_B - V_A}{R_2} + I_2 = 0$$

（4）解方程。

4.6 阅读材料参考译文

4.6.1 网孔（电流）分析法

另一种分析电路的方法是以一组同时的电流为变量，这种方法称为网孔（电流）分析法。对给定一个含有 m 个网孔，不含电流源的平面电路，网孔（电流）分析法的步骤如下：
（1）在 m（有限）个网孔上假设 m 个顺时针的网孔电流。
（2）对每个网孔应用 KVL，先根据网孔电流用欧姆定律求出每个电阻两端的电压，再按顺时针方向求出每个回路的电压代数和并令其等于 0，列出 m 个方程。
（3）解这 m 个方程，求出网孔电流。
以图 4.4 所示的电路为例：
1）设网孔电流 I_1、I_2。
2）应用 KVL，对第一个网孔有

$$I_1 R_1 + (I_1 - I_2) R_3 - 6 = 0$$

对第二个网孔有

$$I_2 R_2 + 8 + (I_2 - I_1) R_3 = 0$$

（4）解方程。

4.6.2 模拟信号

模拟信号是一个时间连续信号，其中信号随时间变化的特点可以用一些其他时间变化量来表示。一个模拟信号借助一些中间的变量来传送信号的信息。在电气电路中，这个变量最常用的是电压，其次是频率、电流和电荷。

任何信息都可以用一个模拟信号来传输，通常这样一个信号是对物理现象如声音、光、温度、位置或压力的变化产生的一个可测的响应，通过一个传感器可获得的相应的模拟信号。

例如，在记录声音时，空气压力（也就是说、声音）的波动撞击在话筒的振动膜上，引起一个电路中的电压或电流的相应的波动。这个电压或电流就是声音的模拟信号。

理论上，任何测到的模拟信号都有噪音且信号变化速率也是受到限制的，所以模拟系统和数字系统在分辨率和频带宽度上都有限制。实际上，一个模拟系统比较复杂，非线性效应和噪音最终会降低模拟信号的分辨率。在模拟系统中，很难发现是否出现这种分辨率的降低。而在数字系统中，不但可以发现这些噪音等干扰，而且能除去这些干扰。

4.6.3 戴维宁定理

设一个负载电阻 R_L 与一个任意电路（这个电路中只含有线性元件）相连接，如图 4.5（a）所示，那么电阻 R_L 取何值时可以从电路中吸收最多的电功率呢？对一个具体的电路，我们可以用节点分析法或网孔分析法来求出 R_L 所吸收功率的表达式。然后根据这个表达式求出电阻为多大值时吸收的功率最大。用这种方法来求的话计算量比较大，（幸亏有）一个著名的、重要的电路理论概念给出：如果只要求 R_L 的话，图 4.5（a）中的任意电路可以用一个独立电压源串联一个电阻来表示［图 4.5（c）］。

只要求出了这个电压源和这个电阻的值，就可以很方便地分析电阻消耗的最大功率问题。

假设给定的任意电路中含有若干个（部分或全部）下列元件：电阻、电压源、电流源（电源可以是受控源也可以是独立源）。在图上指定两个节点，称节点 a 和 b，则电路可以分成如图 4.5（b）所示的两部分。再假设电路 A 部分中没有受电路 B 部分中变量控制的受控电源，反过来也一样（即假设电路 B 部分中没有受电路 A 部分中变量控制的受控电源）。则可用一个适当的独立电压源 u_{OC} 与一个适当的电阻 R_O 串联的电路来取代电路 A 部分，保证其对电路 B 的作用与电路 A（对电路 B 的作用）完全相同。这个电压源和串联电阻的组合称为电路 A 的戴维宁等效电路。换言之，图 4.5（a）中电路 A 和图 4.5（c）中的左框中的电路对电路 B 有相同的作用。这个结果称为戴维宁定理，是电路理论中比较有用和有意义的概念之一。

求开路端电压 u_{OC} 方法是从电路 A 旁边移走电路 B，如图 4.6（a）所示，求出节点 a 和 b 之间的电压，这个电压就是 u_{OC}。

求戴维宁等效电阻或称输出电阻 R_O 的方法也是从电路 A 旁边移走电路 B，并将电路 A 中所有的独立电源除去只留下受控电源（除源即把电路中的电压源等效成短路，电流源等效成开路），求出节点 a 和 b 之间的电阻就是 R_O，如图 4.6（b）所示。

如果电路 A 中不含受控电源，当除去所有的独立电源之后，得到的可能是一个简单的串并联电阻网络，R_O 也可以通过在节点 a 和 b 之间加一个独立电源，求出（两端）电压与（流入）电流之比来求得。这个过程如图 4.6（c）所示，在大部分情况中，加电压 u_O 求出电流 i_O 或者加电流 i_O 求出电压 u_O 都是可以的。

Unit 5 Alternating current

5.1 Text

An alternating current (AC) is an electrical current whose magnitude and direction vary cyclically, as opposed to direct current, whose direction remains constant.

5.1.1 Alternating current

Alternating currents are accompanied (or caused) by alternating voltages. In English the initialism AC is commonly and somewhat confusingly used for both.

Step and impulse functions are useful in determining the responses of circuits when they are first turned on or when sudden or irregular changes occur in the input; this is called transient analysis. However, to see how a circuit responds to a regular or repetitive input-the steady-state analys is the function that is by far the most useful is the sinusoid.

The sine wave is the most common wave in AC and sometimes we refer to sine AC as AC in short. An AC voltage v can be described mathematically as a function of time by the following equation:

$$v(t) = V_{peak} \sin(\omega t)$$

A sine wave, over one cycle (360°) is shown in Fig 5.1. The dashed line represents the root mean square (RMS) value at about $0.707 V_{peak}$. Where

V_{peak} is the peak voltage (unit: volt)

ω is the angular frequency (unit: radians per second)

T is the time to complete one cycle, and is called period(unit: second)

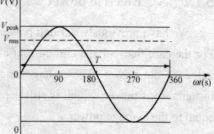

Fig 5.1 sine wave

The angular frequency ω is related to the physical frequency f, which represents the number of oscillations per second (unit: hertz), by the equation $\omega = 2\pi f$.

t is the time (unit: second).

We have following conclusions about the sinusoid:

(1) If the input of a linear, time-invariant circuit is a sinusoid, then the response is sinusoid of the same frequency.

(2) Finding the magnitude and phase angle of a sinusoidal steady-state response can be accomplished with either real or complex sinusoids.

(3) If the output of a sinusoidal circuit reaches its peak before the input, the circuit is a lead network. Conversely, it is a lag network.

(4) Using the concepts of phasor and impedance, sinusoidal circuits can be analyzed in the frequency domain in a manner analogous to resistive circuits by using the phasor versions of KCL, KVL, nodal analysis, mesh analysis and loop analysis.

Though electromechanical generators and many other physical phenomena naturally produce sine waves, this is not the only kind of alternating wave in existence. Other "waveforms" of AC are commonly produced within electronic circuitry. Here are but a few sample waveforms and their common designations in Fig 5.2.

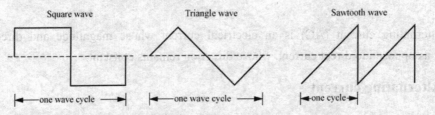

Fig 5.2 some common waveshapes (waveforms)

They're simply a few that are common enough to have been given distinct names. Even in circuits that are supposed to manifest "pure" sine, square, triangle, or sawtooth voltage/current waveforms, the real-life result is often a distorted version of the intended waveshape. Generally speaking, any waveshape bearing close resemblance to a perfect sine wave is termed sinusoidal wave, anything different being labeled as non-sinusoidal wave.

5.1.2 AC Electric power

We have previously defined power to be the product of voltage and current in the DC circuits. For the case that voltage and current are constants, the instantaneous power is equal to the average value of the power. The voltage and current are both sinusoids in AC circuits, however, the instantaneous power, which is still the product of voltage and current, changes with time and is not equal to the average power.

In alternating current circuits, energy storage elements such as inductance and capacitance may result in periodic reversals of the direction of energy flow. The portion of power flow that, averaged over a complete cycle of the AC waveform, results in net transfer of energy in one direction is known as real power (also referred to as active power). That portion of power flow due to stored energy that returns to the source in each cycle is known as reactive power.

The relationship between real power, reactive power and apparent power can be expressed by representing the quantities as vectors. Real power is represented as a horizontal vector and reactive power is represented as a vertical vector (Fig 5.3). The apparent power vector is the hypotenuse of a right triangle formed by connecting the real and reactive power vectors. This

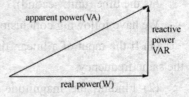

Fig 5.3 power triangle

representation is often called the power triangle. Using the Pythagorean Theorem[1], the relationship among real, reactive and apparent power is:

(apparent power)^2 = (real power)^2 + (reactive power)^2

The ratio of real power to apparent power is called power factor and is a number always between 0 and 1.

5.1.3 Three-Phase Circuits

A circuit that contains a source that produces (sinusoidal) voltages with different phases is called a polyphase system. The importance of this concept lies in the fact that most of he generation and distribution of electric power is accomplished with polyphase systems. The most common polyphase system is the balanced three phase system, which has the property that it supplies constant instantaneous power, which results in less vibration of the rotating machinery used to generate electric power.

A Y(wye)-connected three-phase source in the frequency domain is shown in Fig 5.4, Terminals a, b and c are called the line terminals and n is called the neutral terminal. The source is said to be balanced if the voltages $v_{an}(t)$, $v_{bn}(t)$, $v_{cn}(t)$ (or their phasor $\dot{V}_{an}$, $\dot{V}_{bn}$, $\dot{V}_{cn}$), called phase voltages, have the same amplitude and sum to zero, that is ,if $V_{an}=V_{bn}=V_{cn}$, and $v_{an}(t)+v_{bn}(t)+v_{cn}(t)=0$ (or their phasor form $\dot{V}_{an}+\dot{V}_{bn}+\dot{V}_{cn}=0$).

Fig 5.4 balanced Y-connected three-phase source

Suppose that the amplitude of the sinusoids is V. if we arbitrarily select the angle of $\dot{V}_{an}$ to be aero, that is, if $v_{an}(t)=V\sin(\omega t+0°)$ (or $\dot{V}_{an}=V\angle 0°$), then the two situations that result in a balanced source are as follows:

Case I
$v_{an}(t)=V\sin(\omega t+0°)$
$v_{bn}(t)=V\sin(\omega t-120°)$
$v_{cn}(t)=V\sin(\omega t-240°)$

Case II
$v_{an}(t)=V\sin(\omega t+0°)$
$v_{cn}(t)=V\sin(\omega t-120°)$
$v_{bn}(t)=V\sin(\omega t-240°)$

For case I, $v_{an}(t)$ leads $v_{bn}(t)$ by 120°, and $v_{bn}(t)$ leads $v_{cn}(t)$ by 120°.It is ,therefore, called a positive or abc phase sequence. Similarly, Case II is called a negative or acb phase sequence. Clearly, a negative phase sequence can be converted to a positive phase sequence simply by relabelling the terminals. Thus, we need only consider positive phase sequences.

Let us now connect our balanced source to a balanced Y-connected three-phase load as shown in Fig 5.5, the voltages between the line terminals are called line voltages, the currents between the a and A, b and B, c and C are called line currents. In a balanced Y-connected three-phase load, we have: $V_{line}=\sqrt{3}V_p$ and $I_{line}=I_p$.

If the lines all have the same impedance, the effective load is still balanced and so the neutral current is zero, and the neutral wire can be removed.

More common than a balanced Y-connected three-phase load is a △(delta)-connected load. A Y-connected source with a balanced △-connected load is shown in Fig 5.6 we see that the individual loads are connected directly across the lines, and consequently, it is relatively easier to add or remove one of the components of a △-connected load than with a Y-connected load. In a balanced △-connected three-phase load, we have: $V_{\text{line}} = V_p$ and $I_{\text{line}} = \sqrt{3} I_p$.

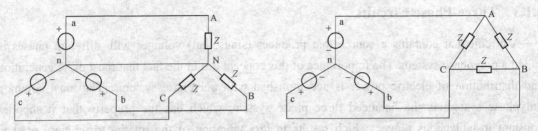

Fig 5.5 balanced Y-Y connected three phase circuit   Fig 5.6 balanced Y-△connected three phase circuits

Regardless of whether a balanced load is Y-connected or △-connected, in terms of the line voltage, line current, and load impedance phase angle, we can use the same formula for the total power absorbed by the load:

$$P = \sqrt{3} V_{\text{line}} I_{\text{line}} \cos\varphi = 3 V_p I_p \cos\varphi$$

Review

(1) An **alternating current** (**AC**) is an electrical current whose magnitude and direction vary cyclically.

(2) The relationship among real, reactive and apparent power is:

$$(\text{apparent power})^2 = (\text{real power})^2 + (\text{reactive power})^2$$

(3) The total power absorbed by the balanced three-phase load:

$$P = \sqrt{3} V_{\text{line}} I_{\text{line}} \cos\varphi = 3 V_p I_p \cos\varphi$$

Note to the Text

[1] Pythagorean Theorem 直角三角形的三条边的关系，以希腊数学家 Pythagorean 命名的。在中国，也早已发现了这个规律，称为勾股弦定律。

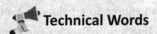 Technical Words

confuse [kənˈfjuːz] vt. 搞乱，使糊涂
cyclical [ˈsiklik(ə)l] adj. 轮转的，循环的 adv. cyclically
dash [dæʃ] n. 破折号，冲撞
distort [diˈstɔːt] vt. 扭曲，失真
hertz [hɜːts] n. 赫，赫兹（频率单位：周/秒）(Hz)

hypotenuse [hai'pɒtənju:z] n. 直角三角形之斜边
initialism [i'nɪʃ(ə)lɪz(ə)m] n. [语]词首字母缩略词
instantaneous [ˌinst(ə)n'teiniəs] adj. 瞬间的，即刻的，即时的
magnitude ['mægnitju:d] n. 大小，数量，幅度
neutral ['nju:tr(əʊ)l] n. 中立者 adj. 中性的
oscillation [ˌɒsɪ'leɪʃn] n. 摆动，振动
period ['pɪərɪəd] n. 时期，周期
phasor ['feɪzə] n. 相位复（数）矢量，相量
polyphase ['pɒlɪfeɪz] adj. 多相的
regardless [rɪ'gɑːdlɪs] adj. 不管，不顾，不注意
relabel [ˌriː'leɪbəl] vt. 重新贴标签于，重新用标签标明
repetitive [rɪ'petɪtɪv] adj. 重复的，反复性的
sawtooth ['sɔːtuːθ] n. 锯齿，锐齿
sequence ['siːkw(ə)ns] n. 次序，顺序，序列
sinusoid ['saɪnəsɔɪd] n. 正弦曲线，正弦
transient ['trænzɪənt] adj. 短暂的，瞬时的 n. 瞬时现象
waveshape ['weɪvʃeɪp] n. 波形

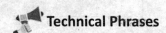

Technical Phrases

peak voltage	峰值电压
linear, time-invariant circuit	线性时不变电路
angular frequency	角频率
root mean square (RMS)	有效值，方均根值
reactive power	无功功率
apparent power	视在功率
balanced three phase system	三相对称系统
transient analysis	暂态分析
positive phase sequence	正相序
△(delta)-connected load	三角形连接的负载

5.2　Reading material

5.2.1　Analysis of AC circuit

Electrical networks that consist only of sources (voltage or current), linear devices (resistors, capacitors, inductors), and linear distributed elements (transmission lines) can be analyzed by algebraic and transform methods to determine DC response, AC response, and transient response.

With the notable exception of calculations for power (P), all AC circuit calculations are based on the same general principles as calculations for DC circuits. The only significant difference is that fact that AC calculations use complex quantities while DC calculations use scalar quantities. Ohm's Law, Kerchief's Laws, and even the network theorems learned in DC still hold true for AC when voltage, current, and impedance are all expressed with complex numbers. The same troubleshooting strategies applied toward DC circuits also hold for AC, although AC can certainly be more difficult to work with due to phase angles.

When faced with analyzing an AC circuit, the first step in analysis is to convert all resistor, inductor, and capacitor component values into impedances (Z), based on the frequency of the power source(Fig 5.7). After that, proceed with the same steps and strategies learned for analyzing DC circuits, using the "new" form of Ohm's Law:

$$\dot{V} = \dot{I}Z, \text{ or } \dot{I} = \dot{V}/Z, \text{ or } Z = \dot{V}/\dot{I}$$

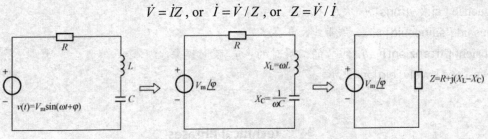

Fig 5.7 AC circuit analysis

5.2.2 Band-pass filter

A band-pass filter is a device that passes frequencies within a certain range and rejects (attenuates) frequencies outside that range (Fig 5.8). An example of an analogue electronic band-pass filter is an RLC circuit (a resistor–inductor–capacitor circuit). These filters can also be created by combining a low-pass filter with a high-pass filter.

Bandpass is an adjective that describes a type of filter or filtering process; it is frequently confused with passband, which refers to the actual portion of affected spectrum. An ideal bandpass filter would have a completely

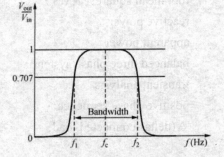

Fig 5.8 an ideal filter passband

flat passband and would completely attenuate all frequencies outside the passband. Additionally, the transition out of the passband would be instantaneous in frequency. In practice, no bandpass filter is ideal. The filter does not attenuate all frequencies outside the desired frequency range completely.

The bandwidth of the filter is simply the difference between the upper and lower cutoff frequencies ($f_2 - f_1$).

5.2.3　Audio Amplifiers

Modest power audio amplifiers for driving small speakers or other light loads can be constructed in a number of ways. The first choice is usually an integrated circuit designed for the purpose. Discrete designs can also be built with readily available transistors or op-amps and many designs are featured in manufacturers' application notes. Older designs employed audio interstage and output transformers but the cost and size of these parts has made them all but disappear. (Actually, when the power source is a 9 volt battery, a push-pull output stage using a 500 Ohm to 8 Ohm transformer is more efficient than non-transformer designs when providing 100 milliwatts of audio.)

Here a simple audio amplifier shows the LM386 in a high-gain configuration (A = 200) (Fig 5.9). For a maximum gain of only 20, leave out the 10 μF connected from pin 1 to pin 8 and a maximum gain of 200 with a 10 μF connected from pin 1 to pin 8. Maximum gains between 20 and 200 may be realized by adding a selected resistor in series with the same 10 μF capacitor. The 10kΩ potentiometer connected with input terminal will give the amplifier a variable gain from zero up to the maximum.

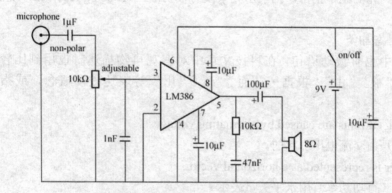

Fig 5.9　a simple audio amplifier

5.3　Knowledge about translation（被动语态）

在科技英语中，经常需要客观地描述一些事物的特点、动作的过程，这时动作的执行者不太清楚，或不太重要或难以明确说出，就经常采用被动语态。

An AC voltage v can be described mathematically as a function of time.

交流电压 v 在数学上可以用一个时间的函数来描述。

句子中 An AC voltage v 是主语，但在被动语态中，主语是动作执行的对象。这里是谁这样去描述交流电压并不重要，或并不想说明，就采用了被动语态。对照一下，也可以这样说：

We can describe mathematically an AC voltage v as a function of time.

我们把交流电压 v 描述成数学上的一个时间函数。

这是一般的语句形式，作为对照，称为主动语态。这里加了动作的执行者 We，作为主语，把 an AC voltage v 作为宾语，有些科普读物中也常加 you，这样比较通俗，但在专业文献中，很少这样用。

1. 被动态的基本表示方法

$$be + V\text{-}ed（及物动词）$$

其中 be 即要表示出与主语的人称和数的一致，也要表示出时态。

Two PN junctions are formed in the device.

器件中形成两个 PN 结。（are formed，主语是复数，一般现在时）

The relationship between real power, reactive power and apparent power can be expressed by representing the quantities as vectors.

有功功率、无功功率和视在功率可以用表示矢量大小（模）的方法来表示。（can be expressed，情态动词后用原形）。

If light is being received at the right frequency then the beam is not broken.

如果接收到光的频率正常，则表示光束没有被切断。（is being received，现在进行时表示的被动语态）。

2. 被动语态翻译

（1）中文中也有被动语句，在科技文章中对客观事物描述时也用到比较多的被动语态，一般用"被""由""根据……而""受""用"等表达被动概念，或指出动作的执行者。

Alternating currents are caused by alternating voltages.

交流电流是由交流电压引起的。

Real power is represented as a horizontal vector.

实数功率（有功功率）用水平矢量表示。

It is said……   据说……

It is reported ……   据报道……

（2）增加"我们""人们""操作者""你"等主语，使句子更容易理解。

Finding the magnitude and phase angle of a sinusoidal steady-state response can be accomplished with either real or complex sinusoids.

通过用实数或复数求解方法，我们可以求出一个正弦稳态响应的幅值和相位角。

（3）可以译成无主语的句子。

In this case, the neutral current is zero, and the neutral wire can be removed.

在这种情况中，中线电流为零，可以不要中线。

Case II is called a negative or acb phase sequence.

第二种情况称为负相序或 acb 相序。

Clearly, faults must be removed from the system as rapidly as possible.

显然，要尽快排除系统故障。

5.4 Exercises

1. Put the Phrases into English (将下列词组译成英语)
(1) 交流电压
(2) 阶跃函数
(3) 正弦波
(4) 稳态分析
(5) 角频率
(6) 幅度和相位
(7) 无功功率
(8) 三相系统
(9) 线电流
(10) 三角形连接

2. Put the Phrases into Chinese (将下列词组译成中文)
(1) impulse function
(2) irregular change
(3) transient analysis
(4) steady-state response
(5) lag network
(6) mesh analysis
(7) sawtooth wave
(8) non-sinusoidal
(9) apparent power
(10) three-phase source

3. Sentence Translation (将下列句子译成中文)

(1) If the output of a sinusoidal circuit reaches its peak before the input, the circuit is a lead network. Conversely, it is a lag network.

(2) A circuit that contains a source that produces (sinusoidal) voltages with different phases is called a polyphase system.

(3) We have previously defined power to be the product of voltage and current in the DC circuits.

(4) The ratio of real power to apparent power is called power factor and is a number always between 0 and 1.

(5) Clearly, a negative phase sequence can be converted to a positive phase sequence simply by relabelling the terminals.

4. Translation (翻译)

Using the MP3 compression system reduces the number of bytes in a song, while retaining sound that is near CD-quality. Consider that an average song is about four minutes long. On a CD, that song uses about 40 megabytes (MB), but uses only 4 MB if compressed through the MP3 format. On average, 64 MB of storage space equals an hour of music. A music listener who has an MP3 player with 1 GB (approximately 1,000 MB) of storage space can carry about 240 songs or the equivalent of about 20 CDs.

5.5 课文参考译文

直流电流的流动方向是保持不变的，交流电与它不同，交流电是指电流的大小和方向都会周期性地变化。

5.5.1 交流电

交流电流总是伴随着交流电压存在（或由交流电压产生的）。在英语中 AC 可以通用的，有时可同时指交流电压和交流电流。

阶跃函数和脉冲函数在分析电路中突然合上电源或断开电源或出现不规则的输入变化时出现的响应是很有用的，这称为暂态分析。但如果想知道电路对一个有规律的或重复的输入信号的响应——稳态分析，到目前为止最有用的函数是正弦函数。

正弦波是交流电最常见的波形，有时我们把正弦交流电简称为交流电。一个交流电压可以用时间的数学函数公式表示：

$$v(t) = V_{\text{peak}} \sin(\omega t)$$

图 5.1 所示为一个周期（360°）正弦波，虚线表示的是有效值，约为峰值电压的 0.707 倍。其中：V_{peak} 是峰值电压（单位：伏特）；ω 是角频率（单位：每秒弧度）；T 是一个完整波形所需的时间，称为周期（单位：秒）；角速度与每秒振荡的次数（单位：赫兹）频率 f 有关，有 $\omega = 2\pi f$；t 是时间（单位：秒）。

对正弦电，我们有如下结论：

（1）如果一个线性，时不变的电路的输入是一个正弦信号，则它的响应是一个同频率的正弦信号。

（2）通过用实数或复数求解我们可以求出一个正弦稳态响应的幅值和相位角。

（3）如果一个正弦电路的输出响应比输入信号先达到峰值，则称此电路是超前网络，否则就称为滞后网络。

（4）用相量和阻抗的概念，在同一频域的正弦电路可以采用类似于分析阻抗电路的方法用基尔霍夫的电流、电压定律，节点电流分析法，网络分析法和回路分析法的相量（复数）形式进行分析。

虽然发电机和许多其他物理现象产生的都是正弦波形交流电，但交流电并不只有这一种形式。在电路中也常产生交流电的其他波形,图 5.2 给出一些波形和它们的名称。

这几个简单的波形常用到,因此有指定的名称。即使在假设出现"纯"的正弦波、方波、三角波和锯齿波电压/电流的电路中,理想波形实际上也经常是变形的波形。一般来说,任何很像正弦波的波形就称为正弦波,其他统称为非正弦波。

5.5.2 交流电功率

以前在直流电路中曾定义功率是电压和电流的乘积,对于电压和电流都是常数的条件下,瞬时功率与平均功率相等。但在交流电路中,电压和电流都是正弦量,瞬时功率仍然是电压和电流的乘积,是随时间变化而变化的,瞬时功率并不等于平均功率。

在交流电路中,存储能量的元件如电感和电容(只能储存或释放能量),会引起能量流作周期性往返流动。能量流中真正(从电源到负载)作单方向传输那一部分,其大小等于交流波形在一个完整的周期中的平均功率,称为实数功率,也称为有功功率。而由于存储元件(储能时)引起的那部分能量流,在每个周期的(放能时)要流回电源的,称为电抗性功率(中文称无功功率)。

有功功率,无功功率和视在功率可以用类似矢量的方法来表示。有功功率可表示成水平矢量,无功功率可表示成垂直矢量(图 5.3),则视在功率矢量就是连接有功功率和无功功率构成的直角三角形的斜边。这种表示方法称为功率三角形。用毕达哥拉斯的理论,有功功率、无功功率和视在功率之间的关系为:

$$(视在功率)^2 = (有功功率)^2 + (无功功率)^2$$

有功功率与视在功率的比值称为功率因数,其值在 0~1 之间。

5.5.3 三相电路

含有不同相位的正弦电压源的电路称为多相系统。因为多数发电机和配电网都是用多相系统实现的,因此这个概念很重要。最常用的多相系统是三相对称系统,它的瞬时输出功率为常数,因此发电时可以减小带动发电机的机器(例如汽轮机,水轮机)的振动。

星形连接的三相电源的连接图如图 5.4 所示,a, b, c 三端称为相线(火线)端,n 端称为中线端。如果相电压 $v_{an}(t), v_{bn}(t), v_{cn}(t)$(或相量 $\dot{V}_{an}, \dot{V}_{bn}, \dot{V}_{cn}$)的幅度相等即 $V_{an}=V_{bn}=V_{cn}$ 且三个相电压之和等于 0 即 $v_{an}(t)+v_{bn}(t)+v_{cn}(t)=0$ 或相量式 $\dot{V}_{an}+\dot{V}_{bn}+\dot{V}_{cn}=0$,则称这个电源为三相对称电源。

设每个电压源的电压幅度为 V,如果我们取 $\dot{V}_{an}$ 的初相位为 0,即 $v_{an}(t)=V\sin(\omega t+0°)$ 或 $\dot{V}_{an}=V\angle 0°$,则一个对称的电源可能出现以下两种可能的情况

第 1 种情况　　　　　　　　　第 2 种情况

$v_{an}(t) = V\sin(\omega t+0°)$　　　　$v_{an}(t) = V\sin(\omega t+0°)$

$v_{bn}(t) = V\sin(\omega t-120°)$　　$v_{cn}(t) = V\sin(\omega t-120°)$

$v_{cn}(t) = V\sin(\omega t-240°)$　　$v_{bn}(t) = V\sin(\omega t-240°)$

对第一种情况中,$v_{an}(t)$ 超前 $v_{bn}(t)$ 120°,而 $v_{bn}(t)$ 超前 $v_{cn}(t)$ 120°,这称为正序或 abc 相序。同样,第二种情况称为逆序或 acb 相序。显然,负序可以通过重新标定终端转换为正序,因此我们只要考虑正序。

现在把一个对称的三相电源连接到一个对称的 Y 连接的三相负载上，如图 5.5 所示。相线（火线）端之间的电压称为线电压，a 和 A，b 和 B，者说 c 和 C 之间的电流称为线电流。对星形连接三相对称负载有：$V_{line} = \sqrt{3}V_p$ 和 $I_{line} = I_p$。

如果线路阻抗相等，有效负载是三相对称的，则中线中的电流等于 0，因此可以不要中线。

与三相对称 Y 接法的负载相比，负载的三角形接法应用更多，一个对称的三相电源连接到一个对称的三角形连接的三相负载上，如图 5.6 所示。可以看到每个负载是直接接到两个相线之间，因此加上或除去一个三角形接法的负载比加上或除去一个星形接法的负载容易。对三相对称的三角形负载有：$V_{line} = V_p$ 和 $I_{line} = \sqrt{3}I_p$。

无论对称负载是星形连接还是三角形连接，根据线电压、线电流和负载的阻抗角，我们可以用同一个公式求出一个三相对称负载吸收的总有功功率：

$$P = \sqrt{3}V_{line}I_{line}\cos\varphi = 3V_pI_p\cos\varphi$$

5.6 阅读材料参考译文

5.6.1 交流电路分析

仅由电源（电压源或电流源），线性器件（电阻、电容、电感）和线性分布的元素（传输线）组成的电路可以用代数和变换的方法来求出直流响应，交流响应和暂态响应。

除了功率的计算比较特别以外，交流电路其他量都是采用与直流电路相同的基本原理进行求解。只是求解时，交流电路中用复数量（相量）进行计算而直流电路中用标量进行计算。在直流电路中导出的欧姆定律、基尔霍夫定律，甚至网络理论对交流电路均适用，只要把电压，电流和复阻抗用复数表示。在直流电路中采用的解题策略对交流电路也适用，虽然交流电路的求解中要用到相位角，求解难度会大一点。

当分析交流电路时，第一步是根据电源的频率把电路中所有的电阻、电感、电容元件值转换成复阻抗（Z）（图 5.7），然后用欧姆定律的"新"形式：$\dot{V} = \dot{I}Z$ 或 $\dot{I} = \dot{V}/Z$ 或 $Z = \dot{V}/\dot{I}$，采用分析直流电路中学到的相同的步骤和方法进行分析。

5.6.2 带通滤波器

带通滤波器是一种器件，它允许一定频率范围内的信号通过，在这个频率范围之外的信号不能（被削弱）通过（图 5.8）。RLC（电阻—电感—电容）电路就是一种模拟带通滤波电路。这种滤波器也可以用一个低通滤波器加一个高通滤波器构成。

带通是一个形容词用来形容一种滤波器或滤波过程，通常会与通带搞混，通带指的是频带中实际受影响的频率范围，一个理想的带通滤波器有一个完全平坦的通带，并能完全削减（滤去）在通带外的频率信号，另外通带（边缘）也是在某个频率点上有一个突变。但实际上没有一个带通滤波器是理想的，实际滤波器并不能完全削弱（滤去）在理想频率范围外的频率信号。

滤波器的带宽是频率上限与频率下限的差值 f_2-f_1。

5.6.3 音频放大器

用来驱动小型扬声器或其他轻型负载的（适当）小功率音频放大器的电路有很多种。首选的通常是用集成电路设计的电路。用晶体管或运算放大器可以很方便地构成分立元件设计的（音频放大器），在很多产品的说明书中都给出了设计电路。比较早的设计是用内部多级音频放大器和输出变压器，但这些部件的成本和体积因数使这些设计被淘汰（实际上当电源电压是9V电池时，推挽输出级用 $500\Omega/8\Omega$ 的变压器效率比不用变压器直接输出100mW的音频信号的效率高）。

图5.9是一个简单的音频放大器，用放大倍数很高（最高可达 $A=200$）的LM386构成。在1脚和8脚之间不接这个 $10\mu F$ 的电容，最大增益只有20，在1脚和8脚之间接一个 $10\mu F$ 的电容，最大增益可达200。如果加一个可调电阻与这个 $10\mu F$ 电容串联，最大的放大倍数就可以在20～200之间调节（图5.9中未接可调电阻，所以最大的放大倍数为200）。调节输入端的 $10k\Omega$ 可调电阻可以使这个放大器的放大倍数在0到其最大放大倍数之间调节。

Unit 6 Digital System

6.1 Text

In digital systems, information is represented by logic variables that can assume values of logic 1 or logic 0. The logic 1 is also called high, true, or on. Logic 0 is also called low, false, or off. Signals in logic systems switch between high and low as the information being represented changes. We often denote logic variables by uppercase letters such as A, B, C.

A single binary digit (0 or 1), called bit, represents a very small amount of information. For example, a logic variable R could be used to represent whether or not it is raining in a particular location (say, R=1 if it is raining, and R=0 if it is not raining).

6.1.1 Digital signal

Digital signals are digital representations of discrete-time signals, which are often derived from analog signals (Fig 6.1).

An analog signal is a datum that changes over time—say, the temperature at a given location; the depth of a certain point in a pond; or the amplitude of the voltage at some node in a circuit—that can be represented as a mathematical function, with time as the free variable and the signal itself as the dependent variable. A discrete-time signal is a sampled version of an analog signal: the value of the datum is noted at fixed intervals (for example, every microsecond) rather than continuously.

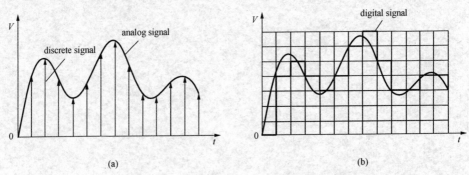

Fig 6.1 digital signal derived from analog signal
(a) sample; (b) quantization

If individual time values of the discrete-time signal, instead of being measured precisely (which would require an infinite number of digits), are approximated to a certain precision—which, therefore, only requires a specific number of digits—then the resultant data stream is termed a digital signal. The process of approximating the precise value within a fixed number of digits, or bits, is called quantization.

In summary, a digital signal is a quantized discrete-time signal; a discrete-time signal is a sampled analog signal.

In the Digital Revolution, the usage of digital signals has increased significantly. Many modern media devices, especially the ones that connect with computers use digital signals to represent signals that were traditionally represented as continuous-time signals; cell phones, music and video players, personal video recorders, and digital cameras are examples. For computers can only "talk" and "think" in terms of binary digital data; A microprocessor can analyse analog data, it must be converted into digital form for the computer to make sense of it.

In most applications, digital signals are represented as binary numbers, so their precision of quantization is measured in bits. Since seven bits, or binary digits, can record 128 discrete values (viz., from 0 to 127), those seven bits are more than sufficient to express a range of one hundred values.

In computer architecture and other digital systems, a waveform that switches between two voltage levels representing a digital signal, even though it is an analog voltage waveform, since it is interpreted in terms of only two levels.

Fig 6.2 Digital signal
①Low level; ②High level;
③Rising edge; ④Falling edge

The clock signal is a special digital signal that is used to synchronize digital circuits. The image shown in Fig 6.2 can be considered the waveform of a clock signal. Logic changes are triggered either by the rising edge or the falling edge.

6.1.2 Gate

The basic circuits used to process the digital signals are gates and flip-flops.

1. The AND Gate

One important logic function is called the AND operation. The AND operation on two logic variables, A and B, is represented as AB, read as "A and B". If the result of A and B is Y, we have the expression: $Y=AB$. A truth table is simply a listing of all of the inputs to a logic operation, together with the corresponding outputs, the truth table for the AND operation of two variables is displayed in Fig 6.3 (a) and circuit symbols for AND gate is showed in Fig 6.3 (b). Notice that AB is 1 only if A and B are both 1 [Fig 6.3 (c)].

2. The NOT Gage

The NOT operation on a logic variable is represented by placing a bar over the symbol for the variable. The symbol $\overline{A}$ is read as "not A" or as "A inverse". If A is 0, $\overline{A}$ is 1, similiarly, if A is 1, $\overline{A}$ is 0.

Circuits that perform the NOT operation are called inverters. The truth table and circuit symbol for an inverter are displayed in Fig 6.4.

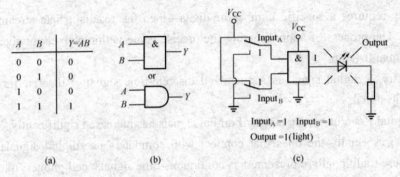

Fig 6.3   AND operation
(a) truth table; (b) circuit symbol; (c) $Y=AB$

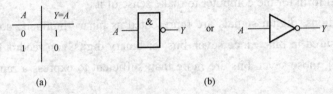

Fig 6.4   NOT operation
(a) truth table; (b) circuit symbol

3. The OR Gate

The OR operation applied to logic variables is written as A+B, which is read as "*A* or *B*". If the result of *A* or *B* is *Y*, we have the expression: $Y=A+B$. The truth table for the OR operation of two variables is displayed in Fig 6.5 (a) and circuit symbols for OR gate is showed in Fig 6.5 (b). Notice that A+B is 1 if A or B (or both) are 1.

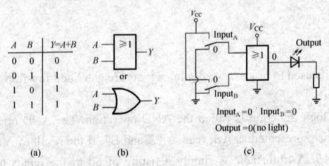

Fig 6.5   OR operation
(a) truth table; (b) circuit symbol; (c) $Y=A+B$

A group of pre-packaged Integrated circuits (ICs) are digital logic circuits. There are several IC families in current use for each of these technologies: transistor-transistor logic (TTL) and complementary metal-oxide semiconductor logic (CMOS). COMS is the most important technology for future systems.

6.1.3 Flip-Flop

In digital circuits, a flip-flop[1] is an onomatopoeic term that refers a kind of bistable multivibrator, an electronic circuit that has two stable states and thereby is capable of serving as one bit of memory.

A flip-flop is usually controlled by one or two control signals and, sometimes, a gate or clock signal. The output often includes the complement ($\bar{Q}$) as well as the normal output (Q). As flip-flops are implemented electronically, they require power and ground connections.

Flip-flops can be either simple (transparent) or clocked. Simple flip-flops can be built by cross-coupling two inverting elements–transistors, NAND gates, or NOR gates (Fig 6.6) – perhaps augmented by some enable/disable (gating) mechanism. RS (reset-set) flip-flop is a simple flip-flop (Fig 6.6) which is the basic part to construct other flip-flops.

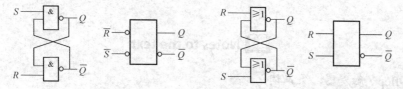

Fig 6.6 Simple flip-flop

Clocked devices are specially designed for synchronous (time-discrete) systems and therefore one such device ignores its inputs except at the transition of a dedicated clock signal (known as clocking, or pulsing). This causes the flip-flop to either change or retain its output signal based upon the values of the input signals at the transition. Some flip-flops change output on the rising edge of the clock, others on the falling edge.

The JK flip-flop is a universal flip-flop, because it can be configured to work as an RS flip-flop, a D flip-flop, or a T flip-flop.

The JK flip-flop's truth table is shown in Fig 6.7(a), and a circuit symbol for a JK flip-flop is shown in Fig 6.7(b), where C is the clock input, J and K are data inputs, Q is the stored data output, and $\bar{Q}$ is the inverse of Q.

Specifically, the combination $J = 1, K = 0$ is a command to set the flip-flop; the combination $J = 0, K = 1$ is a command to reset the flip-flop; and the combination $J = K = 1$ is a command to toggle the flip-flop, i.e., change its output to the logical complement of its current value. Setting $J = K = 0$ will hold the current state. To synthesize a D flip-flop, simply set K equal to the complement of J by connecting a inverter to the input of J. The JK flip flop is positive (rising) edge triggered (Clock Pulse or C in short) as seen in the time diagram in Fig 6.7(c).

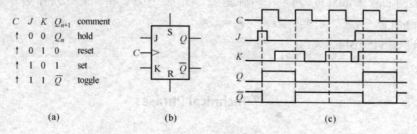

Fig 6.7 JK flip-flop triggered at rising edge
(a) truth table; (b) circuit symbol; (c) the time diagram

Review

(1) In digital systems, information is represented by logic variables that can assume values of logic 1 or logic 0.
(2) A digital signal is a quantized discrete-time signal; a discrete-time signal is a sampled analog signal.
(3) The basic circuits used to process the digital signals are gates and flip-flops.
(4) COMS is the most important technology for future systems.

Notes to the text

[1] flip-flop 触发器，象声词。

Technical Words

amplitude ['æmplitju:d] *n.* 广阔，丰富，振幅
approximate [ə'prɒksimət] *adj.* 近似的，大约的 *v.* 近似，接近，接近，约计
architecture ['ɑ:kitektʃə] *n.* 计算机的物理结构，包括组织结构、容量、该计算机的CPU、存储器以及输入输出设备间的互联
assume [ə'sju:m] *vt.* 假定，设想，采取，呈现
binary ['bainəri] *adj.* 二进制的
complementary [kɒmpli'ment(ə)ri] *adj.* 补充的，补足的
datum ['deitəm] *n.* 数据，资料
denote [di'nəut] *vt.* 指示，表示
interpret [in'tɜ:prit] *v.* 解释，说明，口译
interval ['intəv(ə)l] *n.* 间隔，距离，幕间休息，时间间隔
inverter [in'vɜ:tə] *n.* 反相器
logic ['lɒdʒik] *n.* 逻辑，逻辑学，逻辑性
quantization [ˌkwɒnti'zeʃən] *n.* 量子化 quantize *v.*
synchronize ['siŋkrənaiz] *vi.* 同步，同时发生 *vt.* 使……合拍；使……同步
trigger ['trigə] *vt.* 引发，引起，触发 *n.* 扳机
waveform ['wevfɔrm] *n.* 波形

Technical Phrases

uppercase letters   大写字母
analog signal   模拟信号

discrete-time 离散时间
continuous-time 连续时间
AND Gate 与门
NOT Gate 非门
truth table 真值表

6.2 Reading materials

6.2.1 Analog-to-digital conversion (ADC)

An analog-to-digital converter (abbreviated ADC, A/D or A to D) is an electronic integrated circuit, which converts continuous signals to discrete digital numbers. The reverse operation is performed by a digital-to-analog converter (DAC).

Typically, an ADC is an electronic device that converts an input analog voltage (or current) to a digital number. The digital output may be using different coding schemes, such as binary, Gray code or two's complement binary.

The resolution of the converter indicates the number of discrete values it can produce over the range of analog values. The values are usually stored electronically in binary form, so the resolution is usually expressed in bits. In consequence, the number of discrete values available, or "levels", is usually a power of two. For example, an ADC with a resolution of 8 bits can encode an analog input to one in 256 different levels, since $2^8 = 256$. The values can represent the ranges from 0 to 255 (i.e. unsigned integer) or from -128 to 127 (i.e. signed integer), depending on the application.

Resolution can also be defined electrically, and expressed in volts. The voltage resolution of an ADC is equal to its overall voltage measurement range divided by the number of discrete values as in the formula:

$$Q = \frac{E_{FSR}}{2^M} = \frac{E_{FSR}}{N}$$

Where:

Q is resolution in volts per step (volts per output code),

E_{FSR} is the full scale voltage range

M is the ADC's resolution in bits.

The number of intervals is given by the number of available levels (output codes), which is: $N = 2^M$

Here an example may help:

Full scale measurement range: 0 to 10 volts

ADC resolution is 12 bits: $2^{12} = 4096$ quantization levels (codes)

ADC voltage resolution is: (10V−0V)/4096 codes = 10V/4096 codes ≈ 0.00244 volts/code ≈ 2.44 mV/code

A typical telephone modem makes use of an ADC to convert the incoming audio from a twisted-pair line into signals the computer can understand. In a digital signal processing system, an ADC is required if the signal input is analog.

6.2.2 Integrated circuits

The first integrated circuits contained only a few transistors. Called "Small-Scale Integration" (SSI), they used circuits containing transistors numbering in the tens, and then there are MSI (medium-scale integration), LSI (large-scale integration) and VLSI (very large-scale integration). To reflect further growth of the complexity, the term ULSI that stands for "Ultra-Large Scale Integration" was proposed for chips of complexity of more than 1 million transistors.

System-on-a-Chip (SoC or SOC) is an integrated circuit in which all the components needed for a computer or other system is included on a single chip. The design of such a device can be complex and costly, and building various components on a single piece of silicon may compromise the efficiency of some elements. However, these drawbacks are offset by lower manufacturing and assembly costs and by a greatly reduced power budget: because signals among the components are kept on-die, much less power is required.

Three Dimensional Integrated Circuit (3D-IC) has two or more layers of active electronic components that are integrated both vertically and horizontally into a single circuit. Communication between layers uses on-die signaling, so power consumption is much lower than in equivalent separate circuits. Judicious use of short vertical wires can substantially reduce overall wire length for faster operation.

In the 1980's programmable integrated circuits were developed. These devices contain circuits whose logical function and connectivity can be programmed by the user, rather than being fixed by the integrated circuit manufacturer. This allows a single chip to be programmed to implement different LSI-type functions such as logic gates, adders, and registers. Current devices named FPGAs (Field Programmable Gate Arrays) can now implement tens of thousands of LSI circuits in parallel and operate up to 550 MHz.

6.2.3 555 Timer

The 555 time IC, which is economical and convenient for use in multivibrator circuits because few external components are required. Hence, the device has found wide application. In fact, finding new applications for the 555 has become a game among electronic design engineers. Because of its popularity, several versions of the 555 are available from various manufacturers.

The functional block diagram of the 555 is illustrated in Fig 6.8 , the device contains a resistive voltage-divider string, tow comparators, an RS flip-flop, and a switching transistor. The supply voltage, which can range from 4.5V to 16V, is applied to the series string of three equal resistors. The junction of the top two resistors is externally accessible through the control pin. However, in the applications we consider, the control pin is connected to an open circuit.(Manufacturers of the 555 recommend that a 0.01 μF by-pass capacitor be connected from the control input to ground, preventing power-supply noise from affecting the comparators.) Thus, the voltage divider establishes a voltage of $(2V_{CC})/3$ at the noninverting input of the comparator CP1.

Similarly, the voltage at the inverting input of the comparator CP2 is $V_{CC}/3$.

The reset, threshold, and trigger inputs control the state of the flip-flop. If the reset input is low, the Q output of the flip-flop is low and $\bar{Q}$ is high, current flows into the base of the discharge transistor T; therefore, the transistor T is saturated.

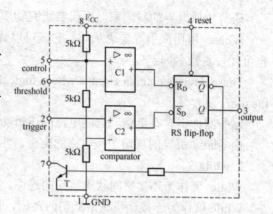

Fig 6.8 block diagram of 555 timer

The reset input has the highest priority in setting the state of the flip-flop. Thus, Q is low if the reset input is low, regardless of the comparator inputs. When the reset input is not in use, it is tied to V_{CC}. and then it does not affect the state of the flip-flop.(Notice that the reset input is active in the low state. Thus, some authors label it as the $\overline{\text{reset}}$ input.)

If the trigger input becomes lower in voltage than the invertting input of CP2 (normally, $V_{CC}/3$), the output of CP2 becomes low, setting the flip-flop. Then Q is high, $\bar{Q}$ is low, and the discharge transistor T is off.(here again, because the trigger input is active in the low state, some writers have labelled it as the $\overline{\text{trigger}}$ input.)

If the threshold input becomes higher in voltage than the noninverting input of CP1 (normally, $2V_{CC}/3$), the output of CP1 becomes low, resetting the flip-flop. Then Q is low, $\bar{Q}$ is high, and the discharge transistor T is in saturation.

6.3　Knowledge about translation（句子的连接Ⅰ）

如果句子只包含一个主谓结构，就称为简单句，如果句子包含两个或更多的主谓结构，就构成并列句或复合句，并列句往往通过并列连词连接，如 and, but 和 or 等，比较容易理解。而复合句一般较长，科技英语中经常用到这样的复杂长句。

复合句中两个或更多的句子中只有一句是主句，其他都称为从句，主句和从句用从属连词连接，在分析时要先找出主句，然后找出从属连词如 that，which，when 等，正确断开句子，把复杂长句分为较简单的句子便于理解。

从属连词只起到连接句子的作用，在从句中不起作用。这里主要介绍从属连词。

1. who, whom, whose

Who(whom)可用作关系代词，引导定语从句，在分析时把从句找出，句子翻译就简单了。
Scientists who deal with the physical universe must deal with both matter and energy.

研究物质世界的科学家，不仅要研究物质，而且要研究能量。

Whose 可用来引导定语从句，它既可以代替人，也可以代替物，既可以代替单数名词，也可代替复数名词。

These devices contain circuits whose logical function and connectivity can be programmed by the user.

这些器件中含有电路，这些电路的逻辑功能和连接可以通过用户编程来改编。

由 whose 引导的从句是作电路的定语。

2. while

While 可作为连接时间状语从句的连词，译成："（正）当……的时候"。

While 可作为连接让步状语从句的连词，译成："虽然，尽管"。

While 还可作为连接两个并列句的连词，译成："而，却，可是"。

Metals in general are good conductors, while nonmetals are insulators.

通常，金属是良导体，而非金属是绝缘体。

3. when

When 引导时间状语从句。

When we talk of electric current, we mean electrons in motion.

当谈论电流时，指的是运动的电子。

When+分词，介词或形容词，这种结构可以看成是省略句，只有从句中的主语和主句中的主语一致而且从句中的谓语一部分是用 be 来表示时，才可以构成这种句型。

When 引导定语从句，用来修饰表示时间概念的名词。

There are times when it is advantageous to substitute one kind of gate for another.

有时可以用一种门来代替另一种门。这里 when 就是 times。

4. After, before, until

after（在……之后），before（在……之前），until (till) 可用作连词，连接时间状语从句。尤其是 until(till)，翻译时注意，若主句是肯定句，可译成"直到……为止"。

CPU instructions reside in memory until required by the CPU.

CPU 指令存储在存储器中直到 CPU 需要时（才调出）。

若主句是否定句，则译成"直到……才"，"在……之前不……"。

The content of the register will be left unchanged until a clock transition applied to the C input of the register.

直到在寄存器的 C 输入端加上一个时钟（脉冲）跃变，寄存器中的内容才会改变。

5. Where

where 在科技英语中常引导地点状语从句和定语从句，Where 也可以引导限定性和非限定性定语从句。

A circuit symbol for a JK flip-flop is shown in Fig 6.7(b), where C is the clock input.

JK 触发器的电路符号如图 6.7（b）所示，其中 C 是时钟（脉冲）输入。

如果 Where 指的不是具体的地点，可译成在……情况下，在……条件下。

6. How、why

科技英语中常用 how 来引导描述操作过程的从句，而 why 则常用来引导说明原因的从句。

The first thing we have to know is how bodies become charged with electricity.
我们必须了解的第一件事，就是物体是怎样带电的。
The question is why metals are good conductors.
问题在于为什么金属是良导体。

7. Because, since, for

这三个词用作连词时所引导的句子都可以表示原因。Because 表示事物本质的原因，直接的理由，事物内在的必然的因果关系；since 只是表示事物内在联系上一种合乎逻辑的自然结果；for 表示间接的、附加的理由，或者是一种推断的理由，并非本质的原因。这三个词，because 语气最强，since 其次，而 for 是并列连词，连接两个并列分句。

6.4 Exercises

1. Put the Phrases into English (将下列词组译成英语)
(1) 数字系统
(2) 逻辑值
(3) 二进制数字
(4) 上升沿
(5) 数学函数
(6) 离散时间信号
(7) 录像机
(8) 真值表
(9) 数字逻辑电路
(10) 用作一位存储器

2. Put the Phrases into Chinese (将下列词组译成中文)
(1) logic variable
(2) whether or not
(3) at fixed time intervals
(4) sampled analog signal
(5) continuous-time signal
(6) be converted into digital form
(7) computer architecture
(8) analog voltage waveform
(9) the corresponding outputs
(10) perform the NOT operation

3. Sentence Translation (将下列句子译成中文)
(1) Signals in logic systems switch between high and low as the information being represented

changes.

(2) Digital signals are digital representations of discrete-time signals, which are often derived from analog signals.

(3) The process of approximating the precise value within a fixed number of digits, or bits, is called quantization.

(4) In most applications, digital signals are represented as binary numbers, so their precision of quantization is measured in bits.

(5) The clock signal is a special digital signal that is used to synchronize digital circuits.

4. Translation (翻译)

An oscilloscope (commonly abbreviated to scope or O-scope) is a type of electronic test equipment(Fig 6.9) that allows signal voltages to be viewed, usually as a two-dimensional graph of one or more electrical potential differences (vertical axis) plotted as a function of time or of some other voltage (horizontal axis).

A typical oscilloscope is a box with a display screen, numerous input connectors, and control knobs and buttons on the front panel. To aid measurement, a grid is drawn on the face of the screen. Each square in the grid is known as a division.

Fig 6.9　oscilloscope

6.5　课文参考译文

在数字系统中，信息是用其值为逻辑 1 或逻辑 0 的逻辑变量来表示的，逻辑值也可以用高（电平）、真或（条件）成立来表示，逻辑 0 也可以用低（电平）、假或（条件）不成立表示。当所表示的信息变化时，逻辑系统的信号在高（电平）和低（电平）之间切换。通常逻辑变量可以用大写字母 A，B，C 表示。

一位二进制数（0 或 1），称为二进制位，表示一个非常小的信息量。例如，逻辑变量 R 可以用来表示在某一个特定的地方是否下雨（即如果下雨，R=1；如果不下雨，R=0）。

6.5.1　数字信号

数字信号是时间离散的信号（即信号不随着时间作连续的变化）的数字表示。数字信号一般是从模拟信号中导出（图 6.1）。

模拟信号是随时间变化的数据流，如在某一指定点的温度，池塘中某指定点的深度或电路中某一节点的电压的值——这些都可以用一个数学函数来表示，其自变量是时间，信号则是一个随时间而变的变量。一个离散时间信号是对模拟信号采样：在每一固定时间间隔（例如每一微秒）的数据值而不是一个连续值。

如果每个离散时间的信号值近似用一个一定精度的值表示而不是精确测量值（精确值可

能需要无限多位数来表示），那么只需要一定的位数就可以表示了——这样表示的数据就称为数字信号。把精确值用一个固定位数的十进制数，或二进制数近似表示的过程称为量化。

总之，一个数字信号是一个量化的离散时间信号，一个离散时间信号是对模拟信号的采样。

在数字化革命中，数字信号的用途有了很大的发展。许多现代的媒体设备，尤其是与计算机相连接的都用数字信号来表示以前用连续时间信号表示的信号，例如话筒、音乐播放器、视频播放器、摄像机和数码相机。因为计算机只能用二进制数据"交流"和"思考"；微处理器可以分析模拟数据，但为了计算机可以了解这些模拟数据的含义必须先把模拟数据转换成数字形式。

在很多应用中，数字信号用二进位制数表示，所以信号的量化精度是用二进制位进行描述的。因为七位二进制数，可以记下 128 个离散的值（从 0～127），这样七位二进制数就足可以表示一个百位数内的数。

在计算机和其他数字系统中，用在两个电平之间变化的波形表示一个数字信号。虽然它是一个模拟电压波形，但它用两个电平来表示数字信号。

时钟信号是一个特定的数字信号，用来同步数字电路（使数字电路中的器件同步动作）。图 6.2 显示的图片可看成是一个时钟信号的波形，在上升沿或下降沿处逻辑电平发生变化。

6.5.2 门（电路）

处理数字信号的基本电路是门电路和触发器。

1. "与"门

一种重要的逻辑函数称为"与"运算，两个逻辑变量 A 和 B 的"与"运算可以用 AB 表示，读作"A 与 B"，如果 A 与 B 是 Y，用表达式 $Y=AB$ 表示。一种逻辑运算可以列出全部的输入情况和相应的输出构成一张真值表。两个变量的"与"运算的真值表如图 6.3（a）所示，"与"门的电路符号如图 6.3（b）所示，注意到仅当 A 和 B 都是 1 时才有 AB=1 [图 6.3（c）]。

2. "非"门

逻辑变量的"非"运算可以在逻辑变量上画一横线来表示，符号 $\overline{A}$ 读作"A 非"或"A 反"，如果 A 等于 0，$\overline{A}$ 等于 1，同样，如果 A=1，则 $\overline{A}$ 等于 0。

完成"非"运算的电路称为反相器，一个反相器的真值表和电路符号如图 6.4 所示。

3. "或"门

加在逻辑变量上的"或"运算写作 $A+B$，可读成"A 或 B"。如果 A 或 B 的结果是 Y，则写出表达式 $Y=A+B$。两个变量的"或"运算的真值表如图 6.5（a）所示，"或"门的逻辑电路符号如图 6.6（b）所示，如果 A 为 1 或 B 为 1 或者 A、B 均为 1 时，$A+B$ 等于 1。

数字逻辑电路有现成封装的集成电路，数字集成电路目前有几个系列，分别采用这样一些技术：三极管—三极管（TTL）逻辑电路和互补型场效应管（CMOS）逻辑电路，CMOS（数字逻辑电路）是未来系统中最重要的技术。

6.5.3 触发器

数字电路中，触发器是一个象声词，指的是一种双稳态多谐振荡器，一种有两个稳定的状态电子电路，触发器因有两个稳定的状态所以可以作为一位存储器。

触发器通常要用一个或两个控制信号进行控制，有时用门电路或时钟（脉冲）信号进行控制。触发器的输出通常有一个输出端（Q），还有一个反相输出端（$\bar{Q}$）。因为触发器的输出是用电信号，因此用触发器时要接上电源和接地。

触发器可以是十分简单（直接通过）或者用时钟脉冲控制的。简单的触发器可以用两个反相元件—晶体管，与非门和或非门交叉耦合组成（图6.6），可能要加上一些使能端。RS触发器是一种基本触发器，是构成其他触发器的基本部件。

受控的触发器可以设计成同步（离散时间）系统，这种器件除非一个指定的时钟信号端（称作时钟，或脉冲）出现跃变，否则无论输入端是什么信号输出都保持不变。当时钟信号端有跃变信号时，触发器根据这一瞬时其输入（端）所加信号的值决定输出是发生变化还是维持原信号。有些触发器是在时钟信号的上升沿改变输出，有些触发器是在时钟信号的下降沿改变输出。

JK触发器是一种通用触发器，因为它可以通过设置实现RS触发器，D触发器和T触发器的功能。

JK触发器的真值表如图6.7（a）所示，图6.7（b）是JK触发器的电路符号，其中C是时钟信号输入端，J和K是数据输入端，Q是储存的数据输出端，$\bar{Q}$是Q的反相输出端。

$J=1$，$K=0$的组合是设置触发器的命令（即触发器的输出$Q=1$），$J=0$，$K=1$的组合是触发器清零的命令（即触发器的输出$Q=0$），$J=1$，$K=1$是触发器触发命令，即输出当前Q值的互补值（即Q取反），设$J=K=0$触发器保持原来的状态（即Q保持不变）。如果要用JK触发器构成D触发器，只要通过在J的输入端接一个反相器，就可以使得K的输入总是等于J输入的互补值。上升沿触发的JK触发器时序图如图6.7（c）所示。

6.6　阅读材料参考译文

6.6.1　模拟/数字转换（ADC）

模拟—数字转换器（缩写成 ADC，A/D）是一个电子集成电路，它把连续信号转换成离散的数字信号。数字—模拟转换器（DAC）的工作过程则相反（把数字信号转换成模拟信号）。

一般情况下，ADC是把输入模拟电压（或电流）转换成数字的电子器件，输入的数字可以用不同的编程方式表示，如二进制数、二—十进制数（格雷编码）或补码形式的二进制数。

转换器的分辨率表示转换器可以在模拟信号的变化范围内产生的离散值的位数，这个值通常用二进制数形式（电子）储存，所以分辨率通常用二进制位数表示，因此，可以转换产生离散值的数目（或称为"级"）通常是2的幂次数。例如，一个8位分辨率的模数转换器可以把一个模拟输入信号范围（从最小到最大）分成256级，因为$2^8 = 256$。根据应用的需要，其值可以表示的范围是0～255（无符号整数），或者从–128～127（即有符号整数）。

分辨率也可以用电量来定义，单位为伏特。一个模数转换器的电压分辨率等于它的电压测量范围除以离散值的数目。如下式：

$$Q = \frac{E_{\text{FSR}}}{2^M} = \frac{E_{\text{FSR}}}{N}$$

其中:
Q 是每级的电压分辨率(每单位输出编码的电压)
E_{FSR} 是全电压(变化)范围,$E_{\text{FSR}}= V_{\text{Hi}}-V_{\text{Lo}}$
M 是模—数转换器以位数表示的分辨率。
间隔的数目是全部级(输出代码)的数目,$N=2^M$。
这里是一个例子以帮助理解:
全测量范围:0~10V。
模—数转换器是 12 位:2^{12} = 4096 量化级(数码)。
模—数转换器电压分辨率是:(10V-0V)/4096=0.00244V/数码≈2.44 mV/数码。
典型的电话调制解调器就是用模数转换器把从双绞线(电话线)送来的音频信号转换成计算机可以处理的(数字)信号,在一个数字处理系统中,输入信号如果是模拟信号,必须要有模数转换器(把它转换成数字信号)。

6.6.2 集成电路

第一块集成电路只含有几个晶体管,称为小规模集成电路(SSI),小规模集成电路一般只含有几十个晶体管。接着就有中规模集成电路(MSI)、大规模集成电路(LSI)和非常大规模集成电路(VLSI)。为了进一步反映集成复杂度的提高,ULSI 这个词用来表示超大规模集成电路,表示含有超过一百万个晶体管的集成电路芯片。

单芯片系统(SoC 或 SOC)是把一个计算机或其他系统所需要的全部器件全集成在单一芯片上的集成电路。这样一个器件的设计比较复杂,设计成本比较高,且在一片硅片上设计各式各样的部件可能会影响其中一些元件的性能。但是这些缺点可以被较低的制造和封装成本,和低功耗(的优点)所弥补。因为器件中的信号都在一个芯片中,只需要极小的功率。

三维集成电路(3D-IC)有两层或更多层有源电子器件,把它们在水平方向和垂直方向都集成在一个电路中。层与层之间的联系用芯片级信号,所以功耗比等效的分立电路要小得多。短竖线的巧妙应用能减少传输线的总长度,且提高运算速度。

在 20 世纪 80 年代开发了可编程的集成电路。其所含的电路的逻辑功能和连接方式不是在制造集成电路时就确定的,而是用户可以通过编程来改编的。这样就可以通过对芯片编程使它实现不同的大规模集成电路的功能,如逻辑门、加法器和寄存器。现在有称为 FPGAs 的器件(可编程门阵列)可以并行实现好几万个大规模集成电路,运算速度达到550MHz。

6.6.3 555 定时器

555 定时集成电路,因为只需要几个外接的器件,就可以构成便宜且很方便地多谐振荡电路。因此 555 定时器的应用很广。实际上对电子设计工程师来说发现 555 定时器的新应用已经成为一种游戏。因为 555 定时器非常普及,很多工厂生产出各种版本的 555 定时器。

555 定时器的功能模块如图 6.8 所示。它含有一个电阻构成的分压器、两个比较器、一个 RS 触发器和一个开关型晶体管。它的电源电压可以在 4.5~16V 的范围内,这个电压加到三个串联的等值电阻上,上面两个电阻的节点引出一个控制脚,可以外接信号,但在所考虑的应用中，控制脚不接信号（555 生产商推荐从控制脚接一个 0.01μF 的旁路电容到地，可以避免电源电压波动噪声影响比较器）。因此分压器在比较器 CP1 的同相输入端加上一个$(2V_{CC})/3$的电压。

同样，在比较器 CP2 的反相输入端的电压是 $V_{CC}/3$。

清零脚，阈值引脚，触发引脚控制触发器的状态，如果清零脚输入为低电平，则触发器的 Q 端输出是低电平，$\bar{Q}$ 是高电平，电流流入放电晶体管 VT 的基极，则晶体管 VT 饱和导通。

在设置触发器的状态时清零脚有最高的优先权，无论比较器的输入是什么，清零脚为低电平输出端 Q 就是低电平。当清零端不用时，要接 V_{CC}，这时它对触发器的状态没有影响（注意到清零端是低电平有效，所以有时标为 $\overline{reset}$）。

如果触发端输入电压低于比较器 CP2 的反相端电压（一般是 $V_{CC}/3$），则 CP2 的输出为低电平，使触发器置 1，则 Q 输出高电平，是低电平。放电晶体管截止（同样，触发端输入是低电平有效，有时也标为 $\overline{trigger}$）。

如果阈值输入端电压高于 CP1 的同相端（一般是 $2V_{CC}/3$），CP1 的输出为低电平，使触发器清零，则 Q 输出低电平，$\bar{Q}$ 输出高电平，放电晶体管 VT 饱和。

Unit 7 Sensors

7.1 Text

In the broadest definition, a sensor is an object whose purpose is to detect events or changes in its environment, and then provide a corresponding output. Sensors may provide various types of output, but typically use electrical or optical signals.

7.1.1 Introduce to sensors

Sensors convert physical phenomena to measurable signals, typically voltages or currents. Consider a simple temperature measuring device, there will be an increase in output voltage proportional to a temperature rise. A computer could measure the voltage, and convert it to a temperature. The basic physical phenomena typically measured with sensors include:

(1) angular or linear position

(2) acceleration

(3) temperature

(4) pressure or flow rates

Sensors are also called transducers. This is because they convert an input phenomenon to an output in a different form. This transformation relies upon a manufactured device with limitations and imperfection. As a result sensor limitations are often characterized with.

Accuracy-This is the maximum difference between the indicated and actual reading. For example, if a sensor reads a force of 100N with a $\pm 1\%$ accuracy, then the force could be anywhere from 99N to 101N.

Resolution-This is the smallest increment that the sensor can detect, this may also be incorporated into the accuracy value. For example if a sensor measures up to 10 centimeter of linear displacements, and it outputs a number between 0 and 100, then the resolution of the device is 0.1 centimeter.

Linearity-In a linear sensor the input phenomenon has a linear relationship with the output signal. In most sensors this is a desirable feature. When the relationship is not linear, the conversion from the sensor output (e.g., voltage) to a calculated quantity (e.g., force) becomes more complex.

7.1.2 Presence detection

There are two basic ways to detect object presence; contact and proximity. Contact implies that there is mechanical contact and a resulting force between the sensor and the object. Proximity indicates that the object is near, but contact is not required.

1. Contact switches

Contact switches (Fig 7.1) are available as normally open and normally closed. Their housings are reinforced so that they can take repeated mechanical forces. These often have rollers and wear pads for the point of contact. Lightweight contact switches can be purchased for less than a dollar, but heavy duty contact switches will have much higher costs. Examples of applications include motion limit switches and part present detectors.

 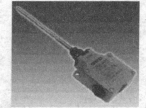

2. Optical (Photoelectric) Sensors

Optical sensors require both a light source (emitter) and detector. Emitters will produce light beams in the visible and invisible spectrums using LEDs and laser diodes. Detectors are typically built with photodiodes or phototransistors. The emitter and detector are positioned so that an object will block or reflect a beam when present. A basic optical sensor is shown in Fig 7.2.

Fig 7.1   contact switches

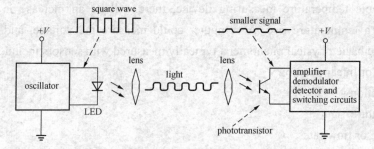

Fig 7.2   a basic optical sensor

In the figure the light beam is generated on the left, focused through a lens. At the detector side the beam is focused on the detector with a second lens. If the beam is broken the detector will indicate an object is present. The oscillating light wave is used so that the sensor can filter out normal light in the room. The light from the emitter is turned on and off at a set frequency. When the detector receives the light it checks to make sure that it is at the same frequency. If light is being received at the right frequency then the beam is not broken. The frequency of oscillation is in the kHz range, and too fast to be noticed. A side effect of the frequency method is that the sensors can be used with lower power at longer distances.

7.1.3 Angular position detection

Potentiometers measure the angular position of a shaft using a variable resistor. A potentiometer is shown in Fig 7.3. The potentiometer is resistor, normally made with a thin film of resistive material. A wiper can be moved along the surface of the resistive film. As the wiper moves toward one end there will be a change in resistance proportional to the distance moved. If a voltage is applied across the resistor, the voltage at the wiper can be detected and the wiper rotates the output voltage will be proportional to the angle of rotation.

Potentiometers are popular because they are inexpensive, and don't require special signal conditioners. But, they have limited accuracy, normally in the range of 1% and they are subject to mechanical wear.

Potentiometers measure absolute position, and they are calibrated by rotating them in their mounting brackets, and then tightening them in place. The range of rotation is normally limited to less than 360 degrees. Some potentiometers can rotate without limits, and the wiper will jump from one end of the resistor to the other.

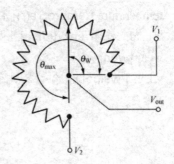

Fig 7.3　principle of a potentiometer

 Review

(1) Sensors convert physical phenomena to measurable signals, typically voltages or currents.
(2) There are two basic ways to detect object presence; contact and proximity.
(3) Optical sensors require both a light source (emitter) and detector.
(4) Potentiometers measure the angular position of a shaft using a variable resistor.

Technical Words

accuracy [ˈækjʊrəsi] n. 精确性，正确度
detector [diˈtektə] n. 发现者，侦察器，探测器，检波器，检电器
imperfection [ˌimpəˈfekʃ(ə)n] n. 不完整性，非理想性，不完美
invisible [inˈvizib(ə)l] adj. 看不见的，无形的
lens [lenz] n. 透镜；镜头，（眼睛的）晶体
linearity [ˌliniˈærəti] n. 线性，直线性
manufacture [mænjuˈfæktʃə] n. 制造者，制造商，制造
maximum [ˈmæksiməm] n. 最大量，最大限度　adj. 最多的，最大极限的
phenomenon [fiˈnɒminən] n. 现象
phototransistor [ˌfəʊtəʊtrænˈzistə] 光电晶体管，光敏晶体（三极）管
potentiometer [pə(ʊ)ˌtenʃiˈɒmitə] n. 电位计，分压计
proximity [prɒkˈsimiti] n. 接近，亲近，近似
purchase [ˈpɜːtʃəs] vt. 买，购买　n. 买，购买
resolution [rezəˈluːʃ(ə)n] n. 分辨率，坚定，决心
roller [ˈrəʊlə] n. 滚筒，辊子
sensor [ˈsensə] n. 传感器

temperature [temprətʃə(r)] *n.* 温度
transducer [trænz'dju:sə] *n.* 传感器，变频器，变换器
wear [weə] *vt.* 穿，戴
wiper ['waipə] *n.* 滑动接触点，滑刷，擦拭者，手帕

Technical Phrases

side effect	副作用，其他作用
physical phenomena	物理现象
contact switches	接触开关
visible spectrum	可见光谱
motion limit switch	运动极限开关，行程开关
wear out	磨损

7.2 Reading material

7.2.1 Inductive Sensors

Inductive sensors use currents induced by magnetic fields to detect nearby metal objects. The inductive sensor uses a coil (an inductor) to generate a high frequency magnetic field as shown in Fig 7.4. If there is a metal object near the changing magnetic field, current will flow in the object. This resulting current flow sets up a new magnetic field that opposes the original magnetic field. The net effect is that it changes the inductance of the coil in the inductive sensor. By measuring the inductance the sensor can determine when a metal have been brought nearby.

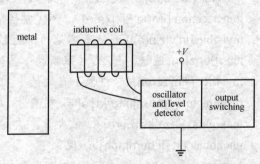

Fig 7.4   principle of an inductive sensor

These sensors will detect any metals, this work by setting up a high frequency field. If a target nears the field will induce eddy currents. These currents consume power because of resistance, so energy is in the field is lost, and the signal amplitude decreases. The detector examines filed magnitude to determine when it has decreased enough to switch.

7.2.2 Encoders

The encoder contains an optical disk with fine windows etched into it, as shown in Fig 7.5,

Light from emitters passes through the openings in the disk to detectors. As the encoder shaft is rotated, the light beams are broken.

There are two fundamental types of encoders; absolute and incremental. An absolute encoder will measure the position of the shaft for a single rotation. The same shaft angle will always produce the same reading. The output is normally a binary or grey code number. An incremental (or relative) encoder will output two pulses that can be used to determine displacement. Logic circuits or software is used to determine the direction of rotation, and count pulses to determine the displacement. The velocity can be determined by measuring the time between pulses. This encoder is often used in the mouse with the roll axis.

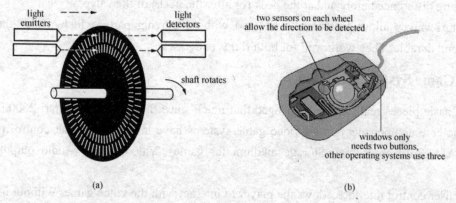

Fig 7.5 encoder
(a) principle of an encoder; (b) a mouse

7.2.3 Velocity detection

1. Tachometers

Tachometers measure the velocity of a rotating shaft. A common technique is to mount a magnet to a rotating shaft. When the magnetic moves past a stationary pick-up coil, current is induced. For each rotation of the shaft there is a pulse in the coil, as shown in Fig 7.6. When the time between the pulses is measured the period for one rotation can be found, and the frequency calculated. This technique often requires some signal conditioning circuitry.

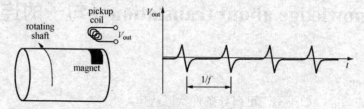

Fig 7.6 a magnetic tachometer

Another common technique uses a simple permanent magnet DC generator (note: you can also use a small DC motor). The generator is hooked to the rotating shaft. The rotation of a shaft will induce a voltage proportional to the angular velocity. This technique will introduce some drag into the system, and is used where efficiency is not an issue.

2. Venturi valve

When a flowing fluid or gas passes through a narrow pipe section (neck) the pressure drops. If there is no flow the pressure before and after the neck will be the same. The faster the fluid flow, the greater the pressure difference before and after the neck. This is known as a Venturi valve. Fig 7.7 shows a Venturi valve being used to measure a fluid flow rate. The fluid flow rate will be proportional to the pressure difference before and at the neck (or after the neck) of the valve.

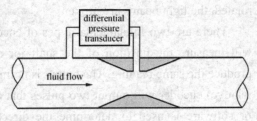

Fig 7.7   venturi valve

Venturi valves allow pressures to be read without moving parts, which makes them very reliable and durable. They work well for both fluids and gases.

7.2.4 Game System Basics

The basic pieces really haven't changed that much since the birth of the Atari 2600. Here's a list of the core components that all video game systems have in common: User control interface, CPU, RAM ,Software kernel ,Storage medium for games ,Video output ,Audio output, Power supply.

The user control interface allows the player to interact with the video game. Without it, a video game would be a passive medium, like cable TV. Early game systems used paddles or joysticks, but most systems today use sophisticated controllers with a variety of buttons and special features.

Ever since the early days of the 2600, video game systems have relied on RAM to provide temporary storage of games as they're being played. Without RAM, even the fastest CPU could not provide the necessary speed for an interactive gaming experience.

The software kernel is the console's operating system. It provides the interface between the various pieces of hardware, allowing the video game programmers to write code using common software libraries and tools.

7.3 Knowledge about translation（句子的连接 II）

1. unless, once

unless 可译成除非，如果不，可引导条件状语从句。
No changes are required in wiring, unless addition of some input or output device is required.
除非需要增加输入或输出，否则不需要改变接线。
once 意为：一旦，一经，只要……便，在……之后，作为连词可引导条件状语从句，或时间条件状语从句。

Once the water level raises enough so that the tank full switch is off (down).

一旦水位升到足够高，水箱满这个开关就断开。

2. Provided (that) 或 providing (that)

连词 provided (that) 意为："如果……，只要……"，可引导条件状语从句。

All substances will permit the passage of some electric current, provided the potential difference is high enough.

如果有足够高的电位差，一切物体都可以传导一些电流。

3. Whether

Whether 作为连词可引导主语从句、表语从句、宾语从句、同位语从句和让步状语从句。尤其是用作宾语从句和让步状语从句用得比较多。译成"是否……"

The substances are divided into two classes, according to whether they did or did not electrify by rubbing.

物质可以根据能否摩擦起电把物质分成两大类。（Whether 引导宾语从句）

A logic variable R could be used to represent whether or not it is raining in a particular location.

用逻辑变量 R 表示在某一个特定的地方是否下雨。（whether or not 引导宾语从句）

4. Although (though)

连词 Although (though) 的意思是"虽然""尽管"，可引导让步状语从句，although 让步语气较重，一般用于正式文章中，though 的让步语气较弱，常用于口语。

People sometimes confuse static electricity with magnetism, though the two are different.

尽管静电和磁是两种不同的现象，但有时人们却把它们二者混为一谈。

5. No matter how (what, where, who, which, when)

其意思是无论怎样（什么、什么地方、谁、哪一个、什么时候）引导让步状语从句。

It is through the electric circuit that energy is transmitted electrically from the primary source, no matter where it is situated to the ultimate consumer.

正是通过电路才能把电能从电源（无论它位于何处）输送到最远的用户。

6. Whenever, wherever, whatever, however

Whenever 意思是"无论何时"，有时也可译成"一……就……""每当……"，"只要……"

Whenever there is a current in a resistor, there is a drop in potential.

只要电阻中有电流，就会有电位降。（这里 whenever 译成"只要……"，"每当……"）同样 Wherever, whatever, however 可分别译成"无论何处，无论什么，无论怎样"。

7.4 Exercise

1. Put the Phrases into English (将下列词组译成英语)
(1) 可测信号
(2) 温度测量器件
(3) 线性传感器

(4) 实际读数

(5) 精确值

(6) 行程开关

(7) 光电传感器

(8) 振荡频率

(9) 绝对位置

(10) 转动的范围

2. Put the Phrases into Chinese

(1) physical phenomena

(2) potentiometer

(3) relies upon

(4) indicated reading

(5) the smallest increment

(6) linear displacement

(7) resolution of the device

(8) linear relationship

(9) output signal

(10) along the surface of the resistive film

3. Sentence Translation (将下列句子译成中文)

(1) Sensors are also called transducers. This is because they convert an input phenomenon to an output in a different form.

(2) In a linear sensor the input phenomenon has a linear relationship with the output signal.

(3) Contact implies that there is mechanical contact and a resulting force between the sensor and the object.

(4) The oscillating light wave is used so that the sensor can filter out normal light in the room.

(5) Potentiometers are popular because they are inexpensive, and don't require special signal conditioners.

4. Translation (翻译)

Most desktop displays use liquid crystal display (LCD) or cathode ray tube (CRT) technology, while nearly all portable computing devices such as laptops incorporate LCD technology. Because of their slimmer design and lower energy consumption, monitors using LCD technology (also called flat panel or flat screen displays) are replacing the venerable CRT on most desktops.

7.5　课文参考译文

在最广泛的定义中，传感器是指一个用来检测的事件或它周围环境的变化，然后提供相应输出信号的器件。传感器可以提供各种类型的输出信号,但最常用的是输出电信号或光信号。

7.5.1 传感器基本概念

传感器把物理现象转换成可测量的信号,一般是电压或电流。先看一个简单的温度测量器件,当温度升高时输出电压会相应的增加。计算机可以测量电压并把它转换成温度(读数显示)。用传感器可以测量的基本物理现象包括:

1) 角度或直线位置。
2) 加速度。
3) 温度。
4) 压力或流量。

传感器也称转换器(中文一般就称传感器),因为它们把输入的物理量转换成不同形式的量输出。因为器件的制造(技术),这种转换(范围)可能有限,(精度等)可能不够。因此传感器通常有以下特征(指标):

精度——指输出指示值和实际值之间的最大差值。例如一个传感器测出一个100N的力,精度为±1%,则实际的力可以是99~101N之间。

分辨率——是传感器可以测到的最小增量,分辨率也可以与精度值合并起来表示。例如传感器最多测到10厘米线性位移,它输出值在0~100之间,则器件的分辨率是0.1cm。

线性度——在一个线性传感器中,输入量与输出信号之间是线性关系。大部分传感器有这个理想的特征。如果不是线性关系,则传感器的输出量(如电压)与计算值(如力)的转换就比较复杂了。

7.5.2 物体测量

测量是否存在一个物体有二种基本方法,接触式和接近式。接触式是指传感器和物体有一个直接接触,在传感器和物体之间产生一个力。接近式是指物体在附近,但并不需要与传感器直接接触。

1. 接触开关

常开和常闭开关可作为接触开关(图 7.1),它们的结构是加强的,所以可以反复承受机械力。在接触点上一般有滚轴或保护垫。轻型接触开关不到1美元就可以买到,但重型的工用接触开关价格比较高。其应用有运动限制和部件到位检测(行程开关)。

2. 光(光电)传感器

光(光电)传感器需要光源(发射器)和检测器,发射器用LED和激光二极管产生可见光或不可见的光束,检测器一般是用光敏二极管或光敏传感器,发射器和检测器是定位放置的,所以有物体时会挡住或反射光束。基本的光学传感器如图7.2所示。

图中左边产生光束,通过透镜聚焦,在检测器一边光速通过第二个透镜聚焦在检测器上。如果光速被切断则检测器检测出有一个物体存在,由于用振荡的光波,所以传感器可滤掉室内正常的照明光。从发射器发出的光按一定频率闪烁,当检测器检测光时确认其按同样的频率闪烁。如果接收到光的频率正常,则表示光束没有被切断。振荡频率在kHz范围,这频率很高所以人不会察觉其闪烁。用振荡的方法还有一个好处是可以用较低功率的传感器进行较远距离光束(检测)。

7.5.3 角度检测

电位计利用可变电阻测量轴转过的角度，图7.3是一个电位计。电位计是一个电阻器，一般是用一层电阻性材料膜做的，一个弧形刷可以在电阻膜的表面移动。当弧形刷向一端运动时电阻值和所移动的距离成正比，如果在电阻两端加上电压，则弧形刷端的电压可以检测出，且弧形刷旋转时输出电压与转过的角度正比。

电位计很便宜，且不需要特别的信号条件，所以应用很广，但它们精度有限，一般在1%左右，并容易磨损。

电位计测量绝对位置，可以把弧形刷转动到它们的支架上，即起始点和最终点进行校正，然后把弧形刷压紧。转动的范围一般小于360°，有些电位计可以没有转动限制，弧形刷会从电阻的一个端点跳到电阻的另一端点上。

7.6 阅读材料参考译文

7.6.1 感应传感器

感应传感器用磁场产生的电流检测附近的金属物体。感应传感器用一个线圈（电感）产生高频磁场如图7.4所示。如果有一个金属物体接近变化的磁场，电流流过物体。这个感应电流产生一个新的电磁场，其作用是反抗原来的电磁场，最终的效果是改变了感应传感器的线圈的电感值，传感器通过测量电感来检测附近是否有金属物体。

这些传感器通过产生一个高频磁场可检测任何金属，如果目标（金属）接近时（金属内）感应产生涡流，因为金属有电阻，这些涡流消耗能量，所以电磁场的能量减小，信号的幅度减小。检测器检测电磁场的幅度，当它小到一定程度时动作。

7.6.2 编码器

编码器含有刻有精细窗孔的透光盘，如图7.5所示。从发射器发出的光通过盘上的开口到达检测器，当编码器的轴旋转时，光速被切断。

有两种基本的编码器，绝对编码器和增量编码器。一个绝对编码器可以测量其轴做单向转动时的位置。相同的轴转角产生相同的读数。输出量一般是二进制数或格雷码。增量（或相对位移）编码器输出两个脉冲，可以用来测量位移，用逻辑电路或软件来计算旋转的方向，由脉冲数求出位移。通过测量脉冲之间的时间还可以求出速度。这种编码器常用在（计算机的）带滚动轴的鼠标中。

7.6.3 速度检测

1. 测速仪

测速仪测量旋转轴的转速，常用的技术是在旋转轴上装一块磁铁，当磁铁转动经过一个静止放置的线圈时，在线圈中产生感应电流。轴每转一圈在线圈中产生一个脉冲，如图7.6

所示，测出脉冲之间的时间就可以求出转动的周期，计算出频率。这种技术一般需要一些信号处理电路。

另一种常用的技术是用一个简单的永磁直流发电机（也可以用一个小的直流电动机）（中文称测速发电机）。发电机与转动轴连接在一起，轴的转动产生一个正比于角速度的电压。这种技术会对原系统的转动产生一些阻碍，在对效率要求不高的系统中可以用。

  2. 文丘里阀

当流动的液体或气体流过一个狭窄的管子颈部压力会下降。如果没有液体流动则颈部前后的相力是相同的。液体流动得越快，则颈部前后的压力差值越大，这就称为文丘里阀。图 7.7 是一个用于测量液体流速率的文丘里阀。液体流动速度是正比于阀的颈部前面和颈部（或颈部后面）的压力差。

文丘里阀没有运动的部件却可以测出压力，这样使得文丘里阀可靠性高，使用时间长。文丘里阀在测量液体和气体流速时工作性能都很好。

7.6.4 （电脑）游戏

自从 Atari 2600（最早的游戏机）问世以来，游戏的基本部分改变得并不多。这里给出所有的视频游戏的核心部分：用户控制界面、CPU、RAM、软件内核、存储游戏的媒体、视频输出、音频输出、电源。

用户控制界面让玩家可以与视频游戏互动，没有它，视频游戏就像有线电视一样成了被动媒体。早期的游戏用简单的手柄，但现在大部分游戏用带有各种按钮和特殊功能的专业控制器。

即使是最早的（2600）游戏机时代，也要靠 RAM 来临时存储所玩的游戏。没有 RAM，即使是最快的 CPU 也不能提供互动游戏所必需的速度。

软件内核是控制台的操作系统，它提供各种硬件之间的接口，使视频游戏程序员可用统一的软件库和工具编写程序代码（设计游戏）。

Unit 8 Electric motor

8.1 Text

An electric motor (Fig 8.1) uses electrical energy to produce mechanical energy. The reverse process, which using mechanical energy to produce electrical energy, is accomplished by a generator or dynamo. Electric motors are found in household appliances such as fans, refrigerators, washing machines, pool pumps, floor vacuum and fan-forced ovens, and many industrial applications are dependent upon motors (or machines) too, which range from the size of one's thumb to the size of a railroad locomotive.

Fig 8.1 various motors

All loads moved by electric motors are really moved by magnetism. The purpose of every component in a motor is to help harness, control, and use magnetic force. To move a load fast does not require more magnets, you just move the magnets fast. To move a heavier load or to decrease acceleration time (accelerate faster), however, more magnets (more torque) are needed. This is the basis for all motor applications.

8.1.1 Induction Motor

The Induction motor is a three phase AC motor and is the most widely used machine. Its characteristic features are:

(1) Simple and rugged construction.
(2) Low cost and minimum maintenance.
(3) High reliability and sufficiently high efficiency.

An Induction motor has basically two parts:

(1) An outside stationary stator having coils supplied with AC current to produce a rotating magnetic field;

(2) An inside rotor attached to the output shaft that is given a torque by the rotating field.

Stator (Fig 8.2)

The Stator is made up of a number of stampings with slots to carry three phase windings. It is wound for a definite number of poles. The windings are geometrically spaced 120 degrees apart.

Stator laminations are stacked together forming a hollow cylinder. Coils of insulated wire are inserted into slots of the stator core.

Each grouping of coils, together with the steel core it surrounds, form an electromagnet. Electromagnetism is the principle behind motor operation. The stator windings are connected directly to the power source.

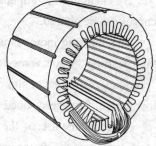

Fig 8.2 stator

Rotor (Fig 8.3)

Two types of rotors are used in Induction motors, squirrel-cage rotor and wound rotor.

The construction of the squirrel cage rotor is reminiscent of rotating exercise wheels found in cages of pet squirrel. The rotor consists of a stack of steel laminations with evenly spaced conductor bars around the circumference.

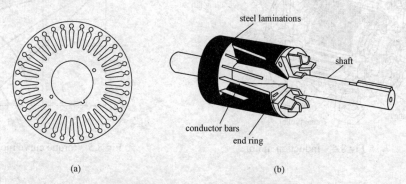

(a) (b)

Fig 8.3 rotor of an induction motor
(a) lamination; (b) the construction of the squirrel cage rotor

The laminations are stacked together to form a rotor core. Aluminum is die cast in the slots of the rotor core to form a series of conductors around the perimeter of the rotor. Current flow through the conductors forms the electromagnet. The conductor bars are mechanically and electrically connected with end rings. The rotor core mounts on a steel shaft to form a rotor assembly.

The enclosure (Fig 8.4) consists of a frame and two end brackets (or bearing housings). The stator is mounted inside the frame. The rotor fits inside the stator with a slight air gap separating it from the stator. There is no direct physical connection between the rotor and the stator. The enclosure also protects the electrical and operating parts of the motor from harmful effects of the environment in which the motor operates. Bearings, mounted on the shaft, support the rotor and allow it to turn. A fan, also mounted on the shaft, is used on the motor for cooling.

The wound rotor motor or slip ring motor is an induction machine where the rotor comprises a set of coils that are terminated in slip rings to which external impedances can be connected. The stator is the same as is used with a standard squirrel cage motor.

By changing the impedance connected to the rotor circuit, the speed/current and speed/torque curves can be altered.

The slip ring motor is used primarily to start a high inertia load or a load that requires a very high starting torque across the full speed range. By correctly selecting the resistors used in the secondary resistance or slip ring starter, the motor is able to produce maximum torque at a relatively low current from zero speed to full speed.

A secondary use of the slip ring motor is to provide a means of speed control. Because the torque curve of the motor is effectively modified by the resistance connected to the rotor circuit, the speed of the motor can be altered (Fig 8.5). Increasing the value of resistance on the rotor circuit will move the speed of maximum torque down. If the resistance connected to the rotor is increased beyond the point where the maximum torque occurs at zero speed, the torque will be further reduced.

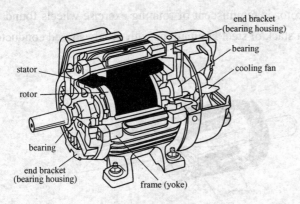

Fig 8.4　induction motor

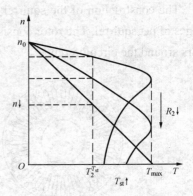

Fig 8.5　torque curve modified by R_2

8.1.2　DC motor

By far the most common DC motor types are the brushed and brushless types, which use internal and external commutation respectively to create an oscillating AC current from the DC source—so they are not purely DC machines in a strict sense.

The classic DC motor design generates an oscillating AC current in a wound rotor with a split ring commutator (Fig 8.6), and either a wound or permanent magnet stator, a rotor consists of a coil wound around a rotor which is then powered by any type of DC voltage source.

Many of the limitations of the classic commutator DC motor are due to the need for brushes to press against the commutator. This creates friction. At higher speeds, brushes have increasing difficulty in maintaining contact. Brushes may bounce off the irregularities in the commutator surface, creating sparks. This limits the maximum speed of the machine. The current density per unit area of the brushes limits the output of the motor. The imperfect electric contact also causes electrical noise. Brushes eventually wear out and require replacement, and the commutator itself is subject to wear and maintenance. The commutator assembly on a large machine is a costly element, requiring precision assembly of many parts.

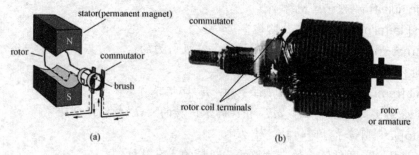

Fig 8.6　DC motor
(a) principle of DC motor; (b) rotor (armature)

Some of the problems of the brushless DC motor are eliminated in the brushless design. In this motor, the mechanical "rotating switch" or commutator/brushgear assembly is replaced by an external electronic switch synchronized to the rotor's position. Brushless motors are typically 85%~90% efficient, whereas DC motors with brush gear are typically 75%~80% efficient.

Brushless DC motors are commonly used where precise speed control is necessary, as in computer disk drives or in video cassette recorders, the spindles within CD drives, and mechanisms within office products such as fans, laser printers and photocopiers. Modern DC brushless motors range in power from a fraction of a watt to many kilowatts. Larger brushless motors up to about 100 kW rating are used in electric vehicles.

 Review

(1) An electric motor uses electrical energy to produce mechanical energy.
(2) An outside stationary stator having coils supplied with AC current to produce a rotating magnetic field;
(3) An inside rotor attached to the output shaft that is given a torque by the rotating field.
(4) A DC motor is designed to run on DC electric power and the most common DC motor types are the brushed and brushless.

Technical Words

acceleration [əkselə'reiʃ(ə)n] n. 加速度
aluminum [ˌæljuˈminiəm] n. 铝
assembly [əˈsembli] n. 集合，装配，集会，集结，汇编，这里指装配组合
circumference [səˈkʌmf(ə)r(ə)ns] n. 圆周，周围
commutator [ˈkɒmjuˌteitə] n. 换向器，转接器

die [dai] n. 骰子，钢型，硬模，冲模
generator ['dʒenəreitə] n. 发电机，发生器
hollow ['hɒləʊ] adj. 空的，凹的 vt. 挖空，弄凹
insulate ['ɪnsjʊleit] v. 使……绝缘
lamination [ˌlæmi'neɪʃən] n. 迭片结构
magnetic [mæg'netik] adj. 磁的，有磁性的，有吸引力的
refrigerator [ri'fridʒəreitə] n. 电冰箱，冷藏库
reminiscent [remi'nis(ə)nt] adj. 回忆往事的 n. 回忆录作者
reverse [ri'vɜːs] n. 相反，背面 adj. 相反的，倒转的 vt. 颠倒，倒转
rotor ['rəʊtə] n. [机]转子，回转轴，转动体
rugged ['rʌgid] adj. 高低不平的，崎岖的，粗糙的，有皱纹的
slot [slɒt] n. 缝，狭槽 vt. 开槽于，跟踪
stationary ['steɪʃ(ə)n(ə)ri] 固定的
stator ['steitə] n. 定子，固定片
torque [tɔːk] n. 扭矩，转矩
winding ['waindiŋ] n. 绕，缠，绕组，线圈

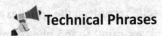

Technical Phrases

squirrel-cage rotor	鼠笼式转子
wound rotor motor	绕线式（异步）电动机
Induction motor	异步电动机，感应电动机
slip ring motor	绕线式（异步）电动机
brushed DC motor	有刷直流电动机
brushless motor	无刷直流电动机

8.2 Reading material

8.2.1 Stepping motors

Stepping motors can be viewed as electric motors without commutators. Typically, all windings in the motor are part of the stator, and the rotor is either a permanent magnet or, in the case of variable reluctance motors, a toothed block of some magnetically soft material.

For example, a variable reluctance stepping motor has three windings, typically connected as shown in the schematic diagram in Fig 8.7, with one terminal common to all windings. In use, the common wire typically goes to the positive supply and the windings are energized in sequence. The rotor in this motor has 4 teeth and the stator has 6 poles, with each winding wrapped around two opposite poles. With winding number 1 energized, the rotor teeth marked X are attracted to this winding's poles. If the current through winding 1 is turned off and winding 2 is turned on, the rotor will rotate 30 degrees clockwise so that the poles marked Y line up with the poles marked 2. To rotate this motor continuously, we just apply power to the 3 windings in sequence.

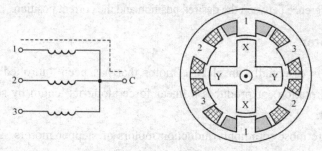

Fig 8.7   variable reluctance motor

All of the commutation must be handled externally by the motor controller, and typically, the motors and controllers are designed so that the motor may be held in any fixed position as well as being rotated one way or the other. Stepping motors come in a wide range of angular resolution. The coarsest motors typically turn 90 degrees per step, while high resolution permanent magnet motors are commonly able to handle 1.8 or even 0.72 degrees per step. With an appropriate controller, most permanent magnet motors can be run in half-steps, and some controllers can handle smaller fractional steps.

Stepping motors can be used in simple open-loop control systems; these are generally adequate for systems that operate at low accelerations with static loads, but closed loop control may be essential for high accelerations, particularly if they involve variable loads.

8.2.2   Servo controllers

Servo controller is a wide category of motor control. Common features are:
(1) precise closed loop position control
(2) fast acceleration rates
(3) precise speed control

Servo motors may be made from several motor types, the most common being:
(1) brushed DC motor
(2) brushless DC motors
(3) AC servo motors

Servo controllers use position feedback to close the control loop. This is commonly implemented with encoders, resolvers, and Hall effect sensors to directly measure the rotor's position. Others measure the back EMF in the undriven coils to infer the rotor position, and therefore are often called "sensorless" controllers.

A servo may be controlled using pulse-width modulation (PWM). How long the pulse remains high (typically between 1 and 2 milliseconds) determines where the motor will try to position itself.

For some applications, there is a choice between using servomotors and stepping motors. Both types of motors offer similar opportunities for precise positioning, but they differ in a number of ways. Servomotors require analog feedback control systems of some type. Typically, this involves a potentiometer to provide feedback about the rotor position, and some mix of circuitry to drive a current through the motor inversely proportional to the difference between the desired position and the current position.

8.2.3 Linear motor

A linear motor is essentially an electric motor that has been "unrolled" so that, instead of producing a torque (rotation), it produces a linear force along its length by setting up a traveling electromagnetic field.

Linear motors are most commonly induction motors or stepper motors. You can find a linear motor in a maglev (Transrapid) train, where the train "flies" over the ground, and in many roller-coasters[1] where the rapid motion of the motorless railcar is controlled by the rail.

Many designs have been put forward for linear motors, falling into two major categories, low-acceleration and high-acceleration linear motors. Low-acceleration linear motors are suitable for maglev trains and other ground-based transportation applications. The Shanghai Maglev Train connects the rapid transit network 30.5 km to the Shanghai Pudong International Airport (Fig 8.8). High-acceleration linear motors are normally quite short, and are designed to accelerate an object up to a very high speed and then release the object, like roller coasters. They are usually used for studies of hypervelocity collisions, as weapons, or as mass drivers for spacecraft propulsion.

The high-acceleration motors are usually of the linear induction design (LIM) with an active three-phase winding on one side of the air-gap and a passive conductor plate on the other side. The low-acceleration, high speed and high power motors are usually of the

Fig 8.8 the Shanghai Maglev Train

linear synchronous design (LSM), with an active winding on one side of the air-gap and an array of alternate-pole magnets on the other side. These magnets can be permanent magnets or energized magnets.

📖 Notes to the reading material

[1] roller-coaster 过山车

8.3 Knowledge about translation（分离现象）

在一般情况下，句中的某些成分应当放在一起，如主语和谓语，动词和宾语等，但在科

技英语中我们常可以看到这些成分离得较远,被其他成分隔开了,这种语言现象叫分离现象。起隔离作用的主要有:

(1) 各种短语:介词短语,分词短语,不定式短语等。

(2) 各种从句。

(3) 句中的附加成分:插入语、同位语和独立成分。

分析这种隔离现象可以更好地帮助我们理解整个句子的原义。常见的分离现象有:

1. 主谓分离

Sometimes certain fields, such as electronic engineering and computer engineering, are considered separate disciplines in their own right.

有时一些领域,例如,电子工程和计算机工程,可以看成是他们各自独立的学科。

The current that should be applied to recharge a 12 V car battery will be very different from the current for a mobile phone battery.

给汽车的 12V 电池充电的电流显然与给手机电池充电的电流大小是不一样的。

2. 动宾分离

有时作状语的介词或介词短语等放在动词之后,用来修饰该动词,而把动词描述动作的对象-宾语隔开了。

A diode is the simplest possible semiconductor device, and is therefore an excellent beginning point if you want to understand how semiconductors work.

二极管是最简单的半导体器件,如果想要了解半导体是如何工作的,最好是从二极管开始。

Note that regardless of the polarity of the input, the current flows in the same direction through the load.

可以看到无论输入电压是正、负极性,流过负载的电流是同样方向的。(把 flows 和 through the load 隔开了。)

3. 复合谓语本身的分离

在复合谓语之间插入含有状语意义的介词短语或状语从句,使复合谓语本身产生分离现象。

In order for a circuit to be properly called electronic, it must contain at least one active device.

如果一个电路被称为电子电路,那么它至少含有一个有源器件。(在这里 be 和 called 复合谓语本身被分离,contain 和 one active device 为动宾分离。)

4. 定语和被修饰名词的分离

An alternative statement of KVL can be obtained by considering voltages across elements **that** are traversed from plus to minus to be positive in sense and voltages across elements **that** are traversed from minus to plus to be negative in sense.

通过假定元件两端的电压是从正到负为正电压,元件两端的电压是从负到正为负电压,可以得到基尔霍夫电压定律另一种描述方法。(这里两个 that 引导的定语从句都修饰 voltages,但被 across elements 隔开了。)

5. 某些词与所要求介词的分离

The electric resistance of a wire is the ratio of the potential difference between its two ends to the current in the wire.

导线的电阻等于该导线两端之间的电位差与导线中电流的比值。[the ratio of A to B, 为

"A"（potential difference）与"B"（the current in the wire）之比，这里被 between its two ends 隔开了。]

8.4　Exercises

1. Put the Phrases into English（将下列词组译成英语）

(1) 异步电动机

(2) 三相交流电动机

(3)（电动机）定子

(4) 旋转（磁）场

(5) 鼠笼式转子

(6) 绕线式电动机

(7) 最大转矩

(8) 在严格意义上

(9) 直流电机

(10) 速度控制

2. Put the Phrases into Chinese（将下列词组译成中文）

(1) to produce mechanical energy

(2) floor vacuum

(3) decrease acceleration time

(4) minimum maintenance

(5) the output shaft

(6) geometrically spaced 120 degrees apart

(7) speed/torque curve

(8) coils of insulated wire

(9) oscillating current

(10) wear out and require replacement

3. Sentence Translation（将下列句子译成中文）

(1) Electric motors are found in household appliances such as fans, refrigerators, washing machines.

(2) To move a load fast does not require more magnets, you just move the magnets fast.

(3) Two types of rotors are used in Induction motors, squirrel-cage rotor and wound rotor.

(4) By changing the impedance connected to the rotor circuit, the speed/current and speed/torque curves can be altered.

(5) The slip ring motor is used primarily to start a high inertia load or a load that requires a very high starting torque across the full speed range.

(6) The imperfect electric contact also causes electrical noise.

(7) At higher speeds, brushes have increasing difficulty in maintaining contact.

(8) In brushless DC motor, the mechanical "rotating switch" or commutator assembly is replaced by an external electronic switch synchronized to the rotor's position.

4. Translation (翻译)

Without a commutator to wear out, the life of a DC brushless motor can be significantly longer compared to a DC motor using brushes and a commutator. Commutation also tends to cause a great deal of electrical and RF noise; without a commutator or brushes, a brushless motor may be used in electrically sensitive devices like audio equipment or computers.

8.5 课文参考译文

电动机（图8.1）是把电能转换成机械能。其逆过程，即把机械能转换成电能是用发电机实现的。在家用电器中常常可以看到电动机，如电风扇、冰箱、洗衣机、游泳池的水泵、地板吸尘器和带鼓风机的炉灶，许多工业电器也要用到电动机，电动机种类很多，小的如拇指般大，大的可以有一节火车厢那么大。

所有用电动机带动的负载实际上是由电磁场驱动的。电动机中的每个部件都是用来处理、控制和利用电磁力的。要让负载（电动机）转得快，并不需要加强电磁场，只需要把电磁场转得快一些（电动机的转速与旋转磁场的转速有关）。但要拖动较重的负载或减少加速的时间（加速度更大），则需要更强的电磁场（输出更大的转矩），这对所有的电动机都是成立的。

8.5.1 感应电动机（异步电动机）

感应电动机是一种三相交流电动机，在各种机器中应用最广。它的主要特征是：

（1）结构简单，坚固。

（2）成本低，维护方便。

（3）可靠性高，效率高。

感应电动机由两部分组成：

（1）（外部）定子——内含定子线圈，用来通入交流电从而产生一个旋转磁场。

（2）（内部）转子——与输出轴相连，在旋转磁场作用下产生转矩。

定子：

定子由许多（冲压的）薄片叠加而成，薄片中间有很多槽（图8.2）用来嵌入三相绕组。槽是根据电动机的磁极对数要求设计的。各相绕组是按在空间互成120°的关系嵌入的。

叠起来并压在一起构成的定子形成一个圆柱形，绝缘的线圈嵌入在定子的槽中。

每组线圈和它周围的铁芯构成一个电磁场，电动机的工作原理就是基于电磁的相互作用，定子线圈直接接到电源上。

转子：

感应电动机的转子有两种类型——鼠笼式转子和绕线式转子（图8.3）。

鼠笼式转子的结构使人想起宠物松鼠笼中的转动的练习轮。转子由绝缘薄钢片叠在一起，周围均匀分布导体棒构成的。

绝缘钢片叠加压制成转子的"核"，表面冲有槽，在槽中是铸铝条，在转子的一周形成一个导体"笼"，流过导体的电流形成电磁场。导体棒在结构上和电路上都是相互连接的。转子"核"装在钢轴上构成转子。

外壳由机架和二端的罩子（或轴承座）组成（图8.4）。定子装在机架中，转子插入定子，定子和转子之间没有直接的连接，相互之间有很小的间隙。当电动机在比较恶劣的环境中工作时，外壳可以保护电动机的电气和运动部分。固定在轴上的后端盖，支撑着转子且让转子可以转动。风扇，也装在轴上，当电动机转动时起散热作用。

绕线式（异步）电动机的转子由转子绕组（和转子铁芯）组成，转子绕组的端子引出，可以外接阻抗。定子与普通鼠笼式电动机的一样。

改变转子电路的阻抗，可以改变速度/电流曲线和速度/转矩曲线（或称机械特性曲线）。

绕线式电动机最常用的场合是带高惯性负载或者需要高起动转矩的负载起动。通过适当切换转子电阻或用绕线式电动机起动器，即使在相当低的工作电流条件下绕线式电动机可以输出其最大转矩，使电动机很快起动（转速从0到额定转速）。

绕线式电动机的第二种用法是可以调速，因为电动机的转矩曲线可以通过改变转子绕组的外接电阻进行调节（图8.5），因而使电动机的速度发生变化。增加转子绕组电阻的值会把出现最大转矩对应的速度向下移。转子电阻到一定值时，在速度为零处且输出最大转矩（以最大转矩起动），如果继续增大转子绕组电阻的值则输出转矩也减小了。

8.5.2 直流电动机

直流电动机是用直流电源供电的，到现在为止最常用的直流电动机有两种：有刷和无刷这是指分别用内部换向器或外接的换向器用输入的直流电源产生一个振荡的交流电流，所以严格来讲，直流电动机并不是一个纯直流的机器。

典型的直流电动机设计，是用一个带有开口的环形换向器在绕线式转子中产生一个振荡电流（图8.6）。无论定子是绕线式的还是永磁式的，转子都是绕线式的，转子线圈接到直流电源上。

许多典型带有换向器的（有刷）直流电动机有一个限制，因为直流电动机运行时需要用电刷压在换向器上转动，这样产生摩擦。速度越高，电刷的接触端维护就越困难。电刷在换向器表面上可能产生不规则的接触（时而接及，时而脱开），会产生火花。因此直流电动机的速度不能太快。直流电动机的输出也因电刷每单位面积流过的电流有限而受到限制。不完善的电接触还会产生电噪声。最终电刷会磨损坏，需要更换，换向器本身也会磨损需要维护。大型直流电动机的换向器装置需要许多部件的精密组合，是成本比较高的。

无刷直流电动机克服了有刷直流电动机的许多问题。在无刷直流电动机中，机械的"旋转开关"即换向器用外接的与转子位置同步的电气开关取代。无刷直流电机的效率可以达到85%～90%，而有刷直流电动机的效率为75%～80%。

无刷直流电动机通常用于需要精确控制速度的场合，如计算机磁盘驱动，盒式录像机中CD的驱动，还有办公室常用的设备如风扇，激光打印机和复印机。现在无刷直流电动机的功率范围很广，小到零点几瓦，大到几千瓦。用在电车中的较大的无刷直流电动机的输出功率可以高达100kW。

8.6 阅读材料参考译文

8.6.1 步进电动机

步进电机可以看成是没有换向器的电动机,通常步进电动机的绕组嵌在定子中,转子齿块则是用永磁材料做的,如果是可变磁阻的步进电动机,其转子齿块是用软磁材料做的。

例如,一个可变磁阻步进电动机有三个线圈,连接如图8.7所示,所有的线圈一端接在一起构成公共端,公共端通常接电源正端,绕组顺序通电,这个电动机的转子有4个齿,定子有6个极,每个绕组绕在相对的两极上。当绕组1通电时,转子的标为X的齿被吸引,与1端对齐,当绕组1断电,绕组2通电时,转子顺时针转过30°,标为Y的齿被吸引,与2端对齐。如果要这个电动机连续转,只要给三个绕时顺序通电。

所有电流的换向是通过电动机控制器在外部实现的。电动机和控制器设计成使电动机可以停在任何固定的位置,也可以按几种模式旋转。根据角度分辨率来区分,步进电动机的种类很多,步子最大的步进电动机每步转过90°,而高分辨率永磁步进电动机每步只转过1.8°甚至0.72°。结合一个适当的控制器,许多永磁步进电动机可以前进半步,有些控制器可以控制步进电动机转过几分之一步。

步进电机可以在简单的开环系统中用,开环系统适用于以低加速度去拖动静止的负载。但对高加速度的操作(即起动快,停止快,也就是响应快),尤其是拖动可变负载时主要是采用闭环控制。

8.6.2 伺服控制器(系统)

伺服控制器(系统)的种类很多,但都有以下共同特点:
(1)精确闭环位置控制。
(2)加速度大。
(3)精确的速度控制。
伺服控制电机有各种电机类型,其中最常用的是:
(1)有刷直流电机。
(2)无刷直流电机。
(3)交流伺服电机。
伺服控制电机是用位置反馈构成闭合控制环,通常是用编码器、求解器和霍尔效应传感器直接测量转子的位置,还有一些伺服控制电机是用通过测量非驱动线圈中的电动势来确定转子的位置,所以常称为"没有传感器"的伺服控制电机。

伺服电机可以用脉宽调制(PWM)来控制,脉冲维持高电平的时间(一般是1~2ms)决定了电动机转过的位置(角度)。

在有些应用场合,可以选择是采用伺服电机还是步进电机。两种电机都可以精确控制。

但有一些地方不同，伺服电机需要某种模拟反馈控制系统，典型的是包含一个指针式仪表给出转子（当前）位置的反馈信号，一些混合电路使得流过电动机的电流与设定的位置和当前位置的差值成反比。

8.6.3　直线（运动）电动机

直线电动机本质上是一个"不转动"的电动机，所以直线电动机不产生一个转矩（旋转），而是通过建立一个前进的电磁场产生沿着长度方向的线性力。

直线电动机最常用的是异步电动机，也有步进电动机。在磁悬浮列车上就用了直线电动机（磁悬浮列车的列车悬在空中），在许多过山车中也用直线电动机通过轨道控制无电动机的过山车厢的快速运动。

直线电动机的设计有很多种，可以归成低加速度和高加速度的直线电动机两大类。低加速度直线电动机适用于磁悬浮列车和其他基于地面的交通。连接上海浦东国际机场的上海磁悬浮列车轨道长 30.5km（图 8.8）。高加速度直线电动机一般相当短，主要设计成加速某一个物体使之达到非常高的速度，然后释放这个物体，像过山车。高加速度直线电动机通常用于高速碰撞的研究，如武器或作航空飞行推进的驱动器。

高加速度直线电动机一般采用直线异步电动机（LIM），在空气间隙的一边带有三相绕组，另一边是无源的导体板。低加速度、高速度、高输出功率的直线电动机一般采用直线同步电动机（LSM），在空气间隙的一边是有源绕组，另一边则是交替排列的磁体。这些磁体可以是永磁体，也可以是励磁产生的磁体。

Unit 9   Motor Controller

9.1   Text

A motor controller is a device or group of devices that serves to govern in some predetermined manner the performance of an electric motor. A motor controller might include a manual or automatic means for starting and stopping the motor, selecting forward or reverse rotation, selecting and regulating the speed, regulating or limiting the torque, and protecting against overloads and faults. A motor controller is connected to a power source such as a battery pack or power supply, and control circuitry in the form of analog or digital input signals.

9.1.1   Solenoid

An electric current through a conductor will produce a magnetic field at right angles to the direction of electron flow. If that conductor is wrapped into a coil shape, the magnetic field produced will be oriented along the length of the coil.(Fig 9.1) The greater the current, the greater the strength of the magnetic field, all other factors being equal.

If we place a magnetic object near such a coil for the purpose of making that object move when we energize the coil with electric current, we have what is called a solenoid. The movable magnetic object is called an armature, and most armatures can be moved with either direct current (DC) or alternating current (AC) energizing the coil. The polarity of the magnetic field is irrelevant for the purpose of attracting an iron armature.

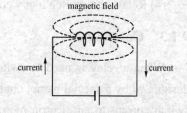

Fig 9.1   direction of the magnetic field

It is noted that a solenoid is a type of electromagnet when the purpose is to generate a controlled magnetic field. If the purpose of the solenoid is instead to impede changes in the electric current, a solenoid can be more specifically classified as an inductor rather than an electromagnet.

Solenoids can be used to electrically open door latches, open or shut valves, move robotic limbs, and even actuate electric switch mechanisms. However, if a solenoid is used to actuate a set of switch contacts, we have a device so useful it deserves its own name: the relay.

9.1.2   Relay

Relays are extremely useful when we have a need to control a large amount of current and/or voltage with a small electrical signal. The relay coil which produces the magnetic field may only consume fractions of a watt of power, while the contacts closed or opened by that magnetic field may be able to conduct hundreds of times that amount of power to a load. In effect, a relay acts as a binary (on or off) amplifier.

In Fig 9.2, the relay's coil is energized by the low-voltage (12 V, DC) source, while the single-pole, single-throw (SPST) contact interrupts the high-voltage (380 V, AC) circuit. It is quite likely that the current required to energize the relay coil will be hundreds of times less than the current rating of the contact. Typical relay coil currents are well below 1 amp, while typical contact ratings for industrial relays are at least 10 amps.

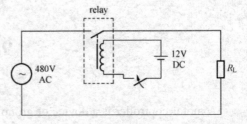

Fig 9.2　load controlled by a relay

One relay coil/armature assembly may be used to actuate more than one set of contacts. Those contacts may be normally-open, normally-closed, or any combination of the two. As with switches, the "normal" state of a relay's contacts is that state when the coil is de-energized, just as you would find the relay sitting on a shelf, not connected to any circuit.

Compare with transistors, relays have following avantages:

(1) Relays can switch AC and DC, transistors can only switch DC.

(2) Relays can switch high voltages, transistors cannot.

(3) Relays are a better choice for switching large currents (> 5A).

(4) Relays can switch many contacts at once.

9.1.3　Contactor

When a relay is used to switch a large amount of electrical power through its contacts, it is designated by a special name: contactor. In other words, a contactor is a large relay, usually used to switch current to an electric motor or other high-power load. Contactors typically have multiple contacts, and those contacts are usually (but not always) normally-open (Fig 9.3), so that power to the load is shut off when the coil is de-energized. Perhaps the most common industrial use for contactors is the control of electric motors.

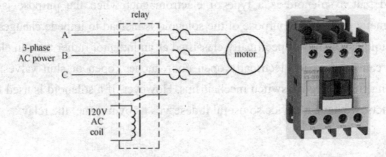

Fig 9.3　AC contactor

The top three contacts switch the respective phases of the incoming 3-phase AC power, typically at least 380 Volts for motors 1 horsepower or greater. The lowest contact is an "auxiliary" contact which has a current rating much lower than that of the large motor power contacts, but is actuated by the same armature as the power contacts.

The auxiliary contact is often used in a relay logic circuit, or for some other part of the motor control scheme, typically switching 120 Volt AC power instead of the motor voltage. One contactor may have several auxiliary contacts, either normally-open or normally-closed, if required.

Large electric motors can be protected from damage through the use of overload heaters and overload contacts of thermal overloads. If the series-connected heaters get too hot from excessive current, the normally-closed overload contact will open, de-energizing the contactor sending power to the motor.

Review

(1) A motor controller serves to govern in some predetermined manner the performance of an electric motor.
(2) Relays control a large amount of current and/or voltage with a small electrical signal.
(3) A contactor is a large relay, usually used to switch current to an electric motor or other high-power load.
(4) Large electric motors can be protected from damage through the use of overload heaters and overload contacts.

Technical Words

actuate [ˈæktʃueit] vt. 开动，促使
armature [ˈɑːmətʃə] n. 衔铁，盔甲，电枢（电机的部件）
auxiliary [ɔːgˈziliəri] adj. 辅助的，补助的
contactor [ˈkɔntæktə] n. 电流接触器
energize [ˈenədʒaiz] n. 给……加电压；供给……能量
interrupt [intəˈrʌpt] vt. 中断，妨碍，插嘴　vi. 打断（别人的讲话或行动）
overcurrent [ˈəuvəkʌrənt] over+current n. 过电流
overload [əuvəˈləud] vt. 使超载，超过负荷　n. 超载，负荷过多
regulate [ˈregjuleit] vt. 管制，控制，调节，校准
relay [ˈriːlei] n. 继电器
scheme [skiːm] n. 安排，配置，计划，图解　v. 计划，设计，图谋，策划
series [ˈsiəriːz] n. 串联，连续，系列，丛书，级数
valve [vælv] n. 阀；活门；气门电子管（心脏）瓣膜

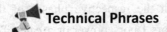 **Technical Phrases**

motor controller	电动机控制器
single-pole, single-throw (SPST) contact	单极单刀接触器
normally-open	动合，常开
normally-closed	动断，常闭
shock absorber	缓冲器，防震器，减震器，阻尼器
sitting on a shelf	现成的（产品）
thermal overload	热继电器
iron armature	衔铁

9.2 Reading materials

9.2.1 Time Relay

Some relays are constructed with a kind of "shock absorber" mechanism attached to the armature which prevents immediate, full motion when the coil is either energized or de-energized. This addition gives the relay the property of time-delay actuation. Time-delay relays can be constructed to delay armature motion on coil energization, de-energization, or both.

Time-delay relay contacts must be specified not only as either normally-open or normally-closed, but whether the delay operates in the direction of closing or in the direction of opening. The following is a description of the four basic types of time-delay relay contacts.

First we have the normally-open, timed-closed (NOTC) contact (Fig 9.4). This type of contact is normally open when the coil is unpowered (de-energized). The contact is closed by the application of power to the relay coil, but only after the coil has been continuously powered for the specified amount of time. In other words, the direction of the contact's motion (either to close or to open) is identical to a regular NO contact, but there is a delay in closing direction. Because the delay occurs in the direction of coil energization, this type of contact is alternatively known as a normally-open, on-delay.

Next we have the normally-open, timed-open (NOTO) contact (Fig 9.5). Like the NOTC contact, this type of contact is normally open when the coil is unpowered (de-energized), and closed by the application of power to the relay coil. However, unlike the NOTC contact, the timing action occurs upon de-energization of the coil rather than upon energization. Because the delay occurs in the direction of coil de-energization, this type of contact is alternatively known as a normally-open, off-delay.

Normally-closed, timed-open. Abbreviated "NCTO", these relays close immediately upon coil de-energization and open only if the coil is continuously energized for the time duration period. Also called normally-closed, on-delay relays.

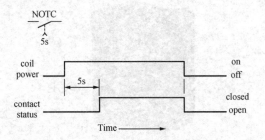

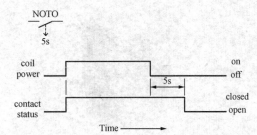

Fig 9.4　timing diagram of NOTC's operation　　Fig 9.5　timing diagram of NOTO's operation

Normally-closed, timed-closed. Abbreviated "NCTC", these relays open immediately upon coil energization and close after the coil has been de-energized for the time duration period. Also called normally-closed, off-delay relays.

9.2.2　Novel controller

Recent developments in drive electronics have allowed efficient and convenient control of these motors, where this has not traditionally been the case. The newest advancements allow for torque generation down to zero speed. This allows the polyphase AC induction motor to compete in areas where DC motors have long dominated, and present an advantage in robustness of design, cost, and reduced maintenance. Here are some examples:

A motor soft starter connects the motor to the power supply through a voltage reduction device and increases the applied voltage gradually or in steps.

QJ Series auto-reduced voltage starter (Fig 9.6) is applicable to infrequent reduced-voltage start up and stop of AC 50Hz three-phase squirrel-cage induction motor with voltage of 380V and power of 10, 14, 28, 40, 55 and 75kW in general industry. When the tapped autotransformer starts, it reduces the power supply voltage, so that the starting current reduces. This product also provides with thermal relay and under-voltage tripping gear. When the motor is overloaded or the voltage of the line is at a certain value below the rated voltage, the motor will be switched off from the power supply to protect the motor.

DK-1B type controller (Fig 9.7) in TKZ series speed-adjustment and control combination device is speed-adjustment motor's direct control unit. It can realize wide-range stepless speed adjustment by operating appointed potentiometer. DK-1B can be equipped with ZC-1A operator to work as single set manual speed adjustment and control or be equipped with DDZ adjustor to work as single set automatic speed adjustment and control.

An Intelligent Motor Controller (IMC) uses a microprocessor to control power electronic devices used for motor control. IMCs monitor the load on a motor and accordingly match motor torque to motor load. This is accomplished by reducing the voltage to the AC terminals and at the same time lowering current and kvar. This can provide a measure of energy efficiency improvement for motors that run under light load for a large part of the time, resulting in less heat, noise, and vibrations generated by the motor.

Fig 9.6   QJ Series auto-reduced voltage starter

Fig 9.7   DK-1B type controller

9.2.3   Experiment material

Relay-contactor Control of three-phase induction motor

I   Objectives

1. To be familiar with relay, contactor, and other controlling devices;
2. To understand the controlling circuits of three-phase induction motors.

II   Apparatus and Equipments

The apparatus and equipments are listed in table 9.1.

Table 9.1   The apparatus and equipments

项目	Name	Spec.	Number
1	multimeter	MF-47	1
2	three-phase induction motor	YS-6314:120W, △/Y, 220/380 V, 0.83/0.48 A	1
3	the experiment board (Fig 9.8)		1
	(1) ac contactor	CJX8-16(B16) (7.5 kW), rated voltage 380 V	2
	(2) thermal overload relay (OLR)	AC.11, the rated current is 2A	1
	(3) button	LAY3	3
	(4) air-breaker Q		1

III   Approach of the Experiment

1. The direct starting of the three-phase induction motor

(1) Be familiar with the structures of devices (see Fig 9.8) and their terminals. Distinguish the normal closed contacts and normal open contacts of the main contactor KM. Write down the data of the induction motor.

(2) Complete following experiment steps after finishing the circuit connection (Fig 9.9).

——Turn on the power supply (air-breaker Q)

——Start the induction motor **directly** (button SB_2)

——Stop the motor by the button SB_1

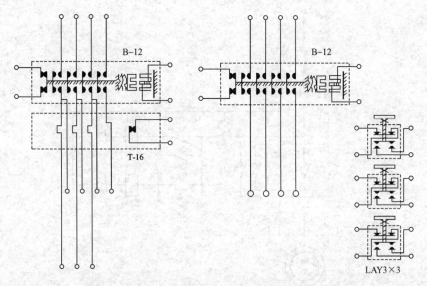

Fig 9.8 Experiment board for relay-contactor control

* **Observe the operation of the contactor KM and the motor**

——Start the motor by button SB_2, then turn off the power supply by the air-breaker Q to stop the motor.

——Turn on the power supply (air-breaker Q) again

* **Observe the operation of the motor and understand low-voltage protection of the circuit.**

——Intermittent operation: turn off the power supply (Q), remove the auxiliary contact connected between the modes a and b, turn on the power supply (Q) again, then operate the button SB_2

* **Understand the function of the auxiliary contact acting as a self-locked contact.**

2. Rotating and inverse rotating of an induction motor

The circuit is shown in Fig 9.10

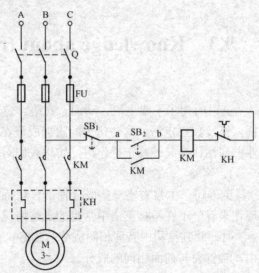

Fig 9.9 the direct starting of the three-phase induction motor

* **Observe the operation of the motor and understand the functions of the auxiliary contacts.**

IV Discussions

1. Point out the protective functions in the circuit shown in Fig 9.10;

2. Point out the self-locked contacts and inter-locked contacts;

3. Whether can the fuse and thermal overload relay replace each other?

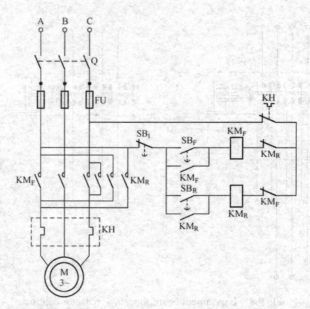

Fig 9.10　Rotating and inverse rotating of an induction motor

9.3　Knowledge about translation（省略和插入语）

英语的句子中除主语、动词、宾语、定语、状语及补足语等主要成分外，还有同位语、插入语及独立定语，这些句子的附加成分在语法中虽然不处于重要的地位，但对阅读科技专业书刊往往会起一些干扰作用，妨碍我们对句子中主要成分的理解和分析。

1. 省略

英语句子中常省略一些成分使语言精练，但有时这种省略给阅读和翻译带来一些困难，现在来看一下在科技英语中常见的省略情况。

（1）并列结构中某些相同成分的省略。在 and，but，or，while，so 等词连起来的并列句中，常省去与前面相同的成分。

Stone, brick and glass conduct heat poorly compared to metals, but well compared to wood, paper, cloth and air. 石头、砖和玻璃与金属相比导热性很差，但与木头、纸、布和空气相比，则导热性较好。（but 后面省去了 they conduct heat）

（2）状语从句中省略。在一些状语从句中，如果其主语与主句中的主语相同，且谓语中含有动词 be，则从句中的主语和 be 常常可以省略。

Consider a simple temperature measuring device, there will be an increase in output voltage proportional to a temperature rise.

先看一个简单的温度测量器件，当温度升高时输出电压会相应的增加。（voltage 后面省去了 which is）

（3）常用省略。有些省略已成为习惯用法，如：if possible（可能的话）if necessary（必要的话），if any（如果主句含有否定的意思，译成：即使有，也很少，如果主句是肯定的，译成：若有……）

In non-conductors (insulators) there are few if any free electrons. 在绝缘体中几乎没有自由电子，即使有，也极少。（这里 if any 也可以不译）

There are very few mistakes in his work, if any. 他工作极少出错。

（4）比较句中的省略。In a series circuit the total resistance is the sum of the individual resistances. The current in each unit is the same as in all other units. 在串联电路中，总电阻等于各个电阻的总和，通过每一个元件的电流与通过其他元件的电流相同。（as 后面省略了 the current is）

2. 插入语

插入语也是英语中常有的现象，它也能造成句子的分离现象，在阅读和翻译时把它先放在一边，免得对句子的主要意思产生干扰。一般插入语会用逗号断开，所以还是比较好判断的。

（1）插入语中见到最多的是同位语，如果名词后面有另一名词（或代词）、词组或句子对该名词作进一步的说明，而这二者在语法上又处于同等的位置，则称为同位语。同位语和前面的名词之间可用标点隔开也可以不隔开。有时用 such as，as，or，that is，for example，particularly（特别是），including（包括）等引出同位语。同位语有时汉语中没有两种说法，可以不译。

Automation **or industrial automation** is the use of control systems **such as computers or PLCs** to control industrial machinery and processes, reducing the need for human intervention.

自动化或工业自动化是用控制系统如计算机或 PLC 去控制工业机器和工作过程，尽量减少人的工作。（用 **or** 和 **such as** 引出同位语）

On October 17,1831, Faraday succeeded in making one of the greatest discoveries in the field of electricity-**the electromagnetic induction**.

法拉第于 1831 年 10 月 7 日成功地做出了电学领域中的一项最伟大的发现——电磁感应。

（2）插入语中常见的还有一些表示语气，态度，承上启下或转折的短语，这样的插入语有许多，典型的如：therefore（所以），for example（例如），for instance（例如），in a conclusion（作为结论），in short（简言之），with……（表示伴随情况）等。但也有的插入语没有用逗号分开，要注意区分：

On multiprocessing operating systems, **however**, a single computer can execute several programs at once.

然而，在多操作系统中，计算机可以同时执行几个程序。

The unit just described, with pulverized coal, air, and water as an input and steam as a useful output, is variously called a steam-generating unit, or furnace, or boiler.

刚才提到的，输入煤粉，空气和水，输出有用的蒸汽的单元有很多个名称，蒸汽发生器，锅炉或沸腾器。（with……，作为插入同位语，在翻译时，为便于理解，译成定语）

9.4　Exercises

1. Put the Phrases into English (将下列词组译成英语)
(1) 电动机控制器
(2) 磁场
(3) 调速
(4) 线圈通电
(5) 关上阀门
(6) 继电器线圈
(7) 动断触点
(8) 辅助触点
(9) 过载保护
(10) 交流接触器

2. Put the Phrases into Chinese (将下列词组译成中文)
(1) starting and stopping the motor
(2) predetermined manner
(3) forward or reverse rotation
(4) digital input signals
(5) be oriented along the length of the coil
(6) attracting an iron armature
(7) industrial relays
(8) sit on a shelf
(9) high-power load
(10) move robotic limbs

3. Sentence Translation (将下列句子译成中文)

(1) A motor controller is connected to a power source such as a battery pack or power supply, and control circuitry in the form of analog or digital input signals.

(2) The greater the current, the greater the strength of the magnetic field, all other factors being equal.

(3) Relays are extremely useful when we have a need to control a large amount of current and/or voltage with a small electrical signal.

(4) It is quite likely that the current required to energize the relay coil will be hundreds of times less than the current rating of the contact.

(5) When a relay is used to switch a large amount of electrical power through its contacts, it is designated by a special name: contactor.

4. Translation (翻译)

(1) Numerical control or numerically controlled (NC) machine tools are machines that are automatically operated by commands encoded on a digital medium. NC machines were first developed soon after World War II and made it possible for large quantities of the desired components to be very precisely and efficiently produced (machined) in a reliable repetitive manner.

(2) 这是一则网上摘下的广告,用词十分简洁。

Apple **iPod nano**

Third Gen. Green (8 GB, MB253LL/A) Digital Media Player

The larger, brighter display means amazing picture quality. It features an anodized aluminum top and polished stainless steel back. It'll wow you for hours. Play up to 5 hours of video or up to 24 hours of audio on a single charge. iTunes provides music, movies, TV shows, games and more. All those features within a wafer-thin, 6.5-mm profile makes **iPod nano** a tiny big attraction you'll just love carrying around!

9.5 课文参考译文

电动机控制器是一个或一组电器,用于按预先设定的方法去控制电动机的运行。一个电动机控制器可以采用手动或自动的手段来起动或制动电动机,选择正转或反转,选择和调节速度,调节或限制电动机的输出转矩,当电动机过载或出现故障时起保护作用。电动机控制器可以直接接在电池组或电源上,用模拟或数字输入信号控制(电动机)电路。

9.5.1 螺旋管

电流流过导体时会产生磁场,磁场的方向和电流的流向符合右手螺旋法则。如果导体绕成一个线圈形状,则所产生磁场的方向沿着线圈的轴线(图9.1)。在其他条件不变的情况下电流越大,则磁场的强度越大。

如果在这个线圈附近放一个铁磁体,当线圈中有电流流过时这个铁磁体就会(受到磁场吸引或排斥而)运动,这个线圈就称为螺旋管,这个运动的铁磁体就称为衔铁。无论在线圈中通上交流电或直流电都可以使这个衔铁运动。磁场的极性与吸引衔铁的目的无关。

应该注意的是,用螺旋管的目的是产生一个可以用电流控制的电磁场,它是一种电磁铁,如果用螺旋管的目的是去抵抗电路中电流的变化,这个螺旋管通常称作(归类成)电感器。

这种螺旋管装置可用于打开门栓,打开或关闭阀门,控制机器人的肢体移动,也可以控制一个电气开关机构。如果这个螺旋管装置用于使一组开关触点动作,就构成了一个十分有用的电器,它有专用的名称:继电器。

9.5.2 继电器

当要用一个小的电信号去控制大电流或高电压时，继电器十分有用。继电器线圈只需要不到1W的功率就可以产生电磁场，由这个电磁场的作用使触点合上或断开，这些触点可以流过很大的电流接通负载（其功率可以是线圈所消耗功率的几百倍）。一个继电器的作用类似于一个二进制量（合上或者断开）放大器。

在图9.2中，继电器的线圈是用低电压（12V直流）电源供电的，而单极性单触点（SPST）则接通或断开高电压（380V交流）电路。给继电器线圈供电的电流往往只是触点额定电流的几百分之一。典型的工业继电器线圈电流远小于1A，而它的触点额定电流至少有10A。

一个继电线的线圈/装置可用来激励不止一组触点，这些触点可以是常开（动合），常闭（动断），或任何两个的组合。就像是开关组，一个继电器的触点的常态是指当线圈不通电，就像你看到的放在架子上没有通电的继电器，其触点的状态。

与晶体管相比，继电器有如下优点：
（1）继电器可以断开或接通交流和直流电路，而晶体管（做开关）只能接通和断开直流电路。
（2）继电路可用于高压电路，晶体管不能用于高压电路。
（3）继电器可用于接通或切断大电流（>5A）。
（4）继电器可同时接通或切断多个触点。

9.5.3 接触器

当继电器用于作为大电流的开关时，就有一个指定的名称：接触器。换言之，接触器是一个大继电器，通常用作电动机或大功率负载的开关切换。典型的接触器有多重触点（一般有三个主触点）（图9.3），这些主触点一般是动合触点（但也有动断触点），所以当线圈中没有电流时电源与负载是不连接的。工业上接触器最主要的用途就是控制电动机。

顶部的三个主触点分别与输入的三相交流电源相应的各相线连接。电动机的电压至少是380V，功率至少是1马力（约为735W）以上。下层的触点是辅助触点，辅助触点的额定电流比用于连接大功率电动机的主触点的额定电流小，主触点和辅助触点都是用相同的（电磁线圈）机构激励的。

通常辅助触点是用在继电逻辑电路或电动机控制电路中，通常额定电压只有120V交流电压，而不是电动机的电压。一个接触器可以有几个辅助触点，有些是动合触点，有些是动断触点。

大电流的电动机还可以用热继电器或过电流接触器进行过（电）流保护。如果串联在主电路中的热继器因为流过的电流过大而过热时，接在控制电路中的热继电器的常闭触点断开，使控制电路失电，从而使接触器的主触点断开，切断电动机的电源。

9.6　阅读材料参考译文

9.6.1　时间继电器

有些继电器的衔铁带有一种阻尼机构，因此当线圈通电或失电时衔铁的动作受到阻尼，不会马上动作。这个附加的阻尼机构使得继电器有延时响应的特点。延时继电器可以设计成通电延时动作或断电延时动作或通电，断电时均延时动作。

使用时间继电器时，不但要指定其触点是动合还是动断，还要指定是通电延时还是失电延时。时间继电器触点有四种基本类型，讨论如下：

（1）先讨论延时闭合动合触点（NOTC）（图9.4）。这类触点在时间继电器不通电时是断开的，当给继电器线圈加上电压时，当连续通电经过一段指定时间后触点闭合。换句话说，触点的断开动作是不延时的，与普通的继电器触点一样，但触点的闭合动作是延时的，因为是通电时延时，这类触点也可以称为通电延时动合触点。

（2）延时断开动合触点（NOTO）（图9.5）。同延时动断常开触点一样，这类触点在时间继电器不通电时是断开的，给继电器线圈加上电压后触点会闭合，但不同的是，在线圈断电时触点延时动作，因为是在线圈断电时动作，这类触点也可以称为断电延时动合触点。

（3）延时断开动断触点（NCTO）。这类触点断电时立即合上，但通电后要延时一段时间才会断开，也称为通电延时动断触点。

（4）延时闭合动断触点（NCTC）。这类触点通电时立即断开，但断电后要延时一段时间才会闭合，也称为断电延时动断触点。

9.6.2 新型控制电器

近年来电气控制方面发展很快，与过去不同的是，现在已经有既方便又高效的电动机的控制方法，最新的技术可以使电动机的输出力矩逐步减小到速度为零。因此在过去只能用直流电动机的场合，现在也可以采用三相交流感应电动机了，因为交流感应电动机在自动化设计、成本和便于维护等方面都（比直流电动机）有优势。以下给出几个例子：

一个电动机软起动器是通过一个降压装置把电动机和电源连接起来，可以（根据电动机降压启动的需要）启动后逐渐加大电动机的电压或分级加大电动机的电压（直到额定电压）。

QJ系列自动减压启动器（图9.6）适用于工业中常用的电压380V，交流50Hz三相鼠笼感应电动机的频繁降压启动和停止（或译制动）。有多种额定功率10、14、28、40、55、75kW类型可以（根据电动机的功率）选择，起动时自动分级接入自耦变压器，降低电压，从而减小启动电流。这个产品还带有热继电器和欠压释放机构（欠压保护装置），当电动机过载或线路电压低于电动机额定电压一定值时，会自动切除电动机的电源以保护电动机。

调速和控制组合电器TKZ系列的DK-1B型控制器（图9.7）是调速电动机的直接控制电器，它可以通过调节电位计实现宽范围的无级调速。DK-1B可以与ZC-1A操作器组合成一台电动机的手动调速控制器；也可以和DDZ调节器组合成一台电动机的自动调速控制器。

智能电动机控制器（IMC）用单片机控制电子功率器件对电动机进行控制。智能电动机控制器监测电动机所带的负载，通过降低交流电源的电压（同时也减小了电流和输入功率）来调节电动机的转矩输出使之与负载匹配。对长时间以轻载运行的电动机来说，可以提高电动机的效率，并减少电动机热耗、噪音和振动。

9.6.3 实验指导书

三相感应电动机的继电器—接触器控制

1. 实验目的

（1）熟悉继电器，接触器和其他控制电器；

（2）理解三相感应电动机的控制电路。
2. 实验设备

实验设备如表 9.1 所列。

表 9.1 实验设备

项目	名称		型号	数量
1	万用表		MF-47	1
2	三相感应电动机		YS-6314:120W，△/Y, 220/380 V, 0.83/0.48A	1
3	实验板（图 9.8）			1
	① 交流接触器		CJX8-16(B16)　(7.5 kW)，额定电压 380 V	2
	② 热继电器 (OLR)		AC.11，额定电流是 2A	1
	③ 按钮		LAY3	3
	④ 空气断路器			1

3. 实验方法

（1）三相感应（异步）电动机的直接起动。

1）熟悉电器的结构和接线端（图 9.8），区分主交流接触器 KM 的常闭触点与常开触点，记下感应（异步）电动机的（铭牌）数据。

2）连接电路（图 9.9）后，按下列实验步骤进行实验：

——合上电源（空气断路器 Q）

——直接起动感应电动机（按钮 SB_2）

——按下按钮 SB_1 停止电动机

*观察接触器 **KM** 和电动机的工作情况。

——按下按钮起动电动机，然后断开空气断路器 Q 切断电源使电动机停止转动。

——再合上空气断路器接通电源。

*观察电动机的工作情况，并理解电路的欠压（失压）保护作用。

——点动运行：断开电源（Q），移去交流接触器辅助触点 a 和 b 的接线，再合上电源（Q），再按下 SB_2

*理解辅助触点的作用是使交流接触器自锁。

（2）三相感应（异步）电动机的正反转运行。

电路如图（9.10）所示。

*观察电动机的运行并理解辅助触点的功能。

4. 讨论

（1）指出图 9.10 中采用的保护措施。

（2）指出自锁和互锁的触点。

（3）熔断器和热继电器是否可以互相替换？

Unit 10　Power System

10.1　Text

Electricity is only one of many forms of energy used in industry, homes, businesses, and transportation. It has many desirable features: it is clean (particularly at the point of use), convenient, relatively easy to transfer from point of source to point of use, and highly flexible in its use. In some cases it is irreplaceable energy source.

The power system consists of power sources, called generating plants (or generators), power end users, called loads, and a transmission and distribution network that connects them. Most commonly the generating plants convert energy from fossil or nuclear fuels, or from falling water, into electrical energy.

10.1.1　Fossil-fuel plant

In a fossil-fuel plant, coal, oil, or natural gas is burned in a furnace. The combustion produced hot water, which is converted to steam, and the steam drives a turbine, which is mechanically coupled to an electric generator. A schematic diagram of a typical coal-fired plant is shown in Fig10.1.

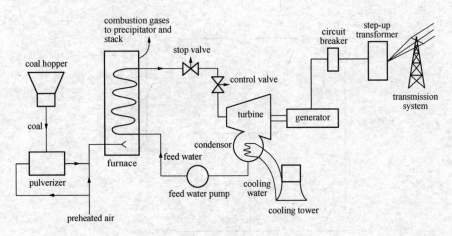

Fig 10.1　coal-fired power plant

In brief, the operation of the plant is as follows: Coal is taken from storage and fed to a pulverizer (or mill), mixed with preheated air, and blown into the furnace, where it is burned.

The furnace contains a complex of tubes and drums, called a boiler, through which water is pumped and the temperature of the water rises in the process until the water evaporates into steam. The steam passes on to the turbine, while the combustion gases (flue gases) are passed through mechanical and electrostatic precipitators, which remove upward of 99% of the solid particles (ash) before being released to the chimney or stack.

The unit just described, with pulverized coal, air, and water as an input and steam as a useful output, is variously called a steam-generating unit, or furnace, or boiler. When the combustion process is under consideration, the term furnace is usually used, while the term boiler is more frequently used when the water-steam cycle is under consideration. The steam, at a typical pressure of 3500 psia[1] and a temperature of 1050 F[2], is supplied through control and stop (shutoff) valves to the steam turbine. The control valve permits the output of the turbine-generator unit (or turbogenerator) to be varied by adjusting steam flow. The stop valve has a protective function.

The turbine turns the rotors of the electric generator in whose stator are embedded three (phase) windings. In the process mechanical power from the turbine drive is converted to three-phase alternating current at voltages in the range from 11 to 30kV line to line at a frequency of 60Hz in the United States. The voltage is usually "stepped up" by transformers for efficient transmission to remote load centres.

10.1.2 Transmission & distribution system

The sources of electric power described in the preceding sections are usually interconnected by a transmission system or network that distributes the power to the various load points or load centres. A small portion of a transmission system that suggests the interconnections is shown as a one-line diagram in Fig 10.2.

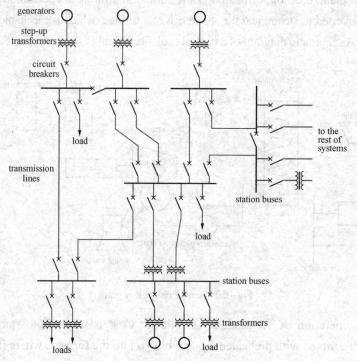

Fig 10.2 small portion of a transmission system

The generator voltages are in the range of 11 to 30kV, higher generator voltages are difficult to obtain because of insulation problems in the narrow confines of the generator stator. Transformers are then used to step up the voltages to the range of 110-765kV for transmission. The voltages refer to voltages from line to line. Near the loads, the transmission voltage is stepped down to the voltages used by loads.

The loads referred to here represent bulk loads, such as the distribution system of a town, city, or large industrial plant. Such distribution systems provide power at various voltage levels. Large industrial consumers or railroads might accept power directly at voltage levels of 23 to 138kV; they would then step down the voltages further. Smaller industrial of commercial consumers typically accept power at voltage levels of 4.16kV to 34.5kV. Residential consumers normally receive single phase power from pole-mounted distribution transformers at voltage levels of 120/240V.

Transmission circuits may be built either underground or overhead (Fig 10.3). Underground cables are used mostly in urban areas where acquisition of overhead rights of way is costly or not possible. They are also used for transmission under rivers, lakes and bays. Overhead transmission is used otherwise because, for a given voltage level, overhead conductors are much less expensive than underground cables.

The distribution system transports the power from the transmission system to the customer. The distribution systems are typically radial because networked systems are more expensive. The equipment associated with the distribution system includes the substation transformers

Fig 10.3 overhead transmission line

connected to the transmission systems, the distribution lines from the transformers to the customers and the protection and control equipment between the transformer and the customer. The protection equipment includes lightning protectors, circuit breakers and fuses. The control equipment includes voltage regulators, capacitors, and demand side management equipment.

One reason for using high transmission-line voltages is to improve energy transmission efficiency. Since the power transmitted is equal to the product of the current, the voltage and the cosine of the phase difference φ ($P = IV\cos\varphi$), the same amount of power can be transmitted with a lower current by increasing the voltage. Why a lower current is needed for transmission? The fact is that the power losses in a conductor are a product of the square of the current and the resistance of the conductor, described by the formula $P=I^2R$. This means that when transmitting a fixed power on a given wire, if the current is doubled, the power loss will be four times greater. Therefore it is advantageous when transmitting large amounts of power to distribute the power with high voltages.

Another reason for higher voltages is the enhancement of stability.

10.1.3   AC power supply frequency

The frequency of the electrical system varies by country; most electric power is generated at either 50 Hz or 60 Hz. Some countries have a mixture of 50 Hz and 60 Hz supplies, notably Japan.

A low frequency eases the design of low speed electric motors, particularly for hoisting, crushing and rolling applications, and commutator-type traction motors for applications such as railways, but also causes a noticeable flicker in incandescent lighting and objectionable flicker of fluorescent lamps. 16.6 Hz power is still used in some European rail systems, such as in Austria, Germany, Norway, Sweden and Switzerland. The use of lower frequencies also provided the advantage of lower impedance losses, which are proportional to frequency. The original Niagara Falls[3] generators were built to produce 25 Hz power, as a compromise between low frequency for traction and heavy induction motors, while still allowing incandescent lighting to operate (although with noticeable flicker); most of the 25 Hz residential and commercial customers for Niagara Falls power were converted to 60 Hz by the late 1950's, although some 25 Hz industrial customers still existed as of the start of the 21st century.

Military, textile industry, marine, computer mainframe, aircraft, and spacecraft applications sometimes use 400 Hz, for benefits of reduced weight of apparatus or higher motor speeds.

Review

(1) The generating plants convert energy from fossil or nuclear fuels, or from falling water, into electrical energy.

(2) The power system consists of generating plants, loads, and a transmission and distribution network.

(3) The frequency of the electrical system varies by country; most electric power is generated at either 50 Hz or 60 Hz.

Notes to the text

[1] psia-绝对压力单位，1psia= 0.0689475728 bar（一个标准大气压）。

[2] F-华氏温标是经验温标之一。在美国的日常生活中，多采用这种温标。与中国习惯的摄氏度的关系为：℃=5/9(F-32)。

[3] Niagara Falls-加拿大的尼亚加拉瀑布。

Technical Words

apparatus [ˌæpəˈreɪtəs] *n.* 器械，设备，仪器

distribution [distriˈbjuːʃ(ə)n] *n.* 分配，分发，配电，发送，发行

drum [drʌm] *n.* 鼓，鼓声，鼓状物 *vt.* 击鼓；大力争取 *vi.* 击鼓

embed [imˈbed] *vt.* 使插入，使嵌入，深留，嵌入

evaporate [iˈvæpəreɪt] *v.* (使)蒸发，消失

flicker [ˈflikə] *n.* 闪烁，闪光 *vi.* 闪动 *vt.* 使摇曳，使闪烁

fossil ['fɒs(ə)l] n. 化石　adj. 化石的，僵化的，守旧的
furnace ['fɜːnɪs] n. 炉子，熔炉
incandescent [ɪnkæn'des(ə)nt] adj. 遇热发光的，白炽的，炽热的
marine [mə'riːn] n. 舰队，水兵，海运业　adj. 海的，海产的，航海的
mill [mɪl] n. 工厂，制造厂，压榨机，磨坊，磨粉机
precipitator [prɪ'sɪpɪteɪtə] n. 催促的人，沉淀剂，催化剂
pulverizer ['pʌlvə,raɪzə] n. 粉碎机，喷雾器
radial ['reɪdɪəl] adj. 光线的，放射状的，半径的　n. 光线，射线
textile ['tekstaɪl] n. 纺织品　adj. 纺织的
transformer [træns'fɔːmə] n. 变压器（变换器，互感器）
transmission [trænz'mɪʃ(ə)n] n. 传动；传送 播送消息 动力传送器
turbine ['tɜːbaɪn] n. 涡轮
upward ['ʌpwəd] adj. 向上的，（声调）声高的　adv. 以上

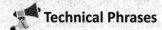

Technical Phrases

circuit breaker	电路断路器
desirable feature	理想的特征
power system	电力系统
transmission network	传输网
distribution network	配电网
fossil-fuel	燃料
lightning protector	避雷器
power loss	功率损失

10.2　Reading materials

10.2.1　Transformer

A transformer(Fig 10.4) is a device that transfers electrical energy from one circuit to another through inductively coupled electrical conductors. A changing current in the first circuit (the primary) creates a changing magnetic field; in turn, this magnetic field induces a changing voltage in the second circuit (the secondary). By adding a load to the secondary circuit, one can make current flow in the transformer, thus transferring energy from one circuit to the other.

The secondary induced voltage V_S, of an ideal transformer, is scaled from the primary V_P by a factor equal to the ratio of the number of turns of wire in their respective windings:

$$\frac{V_S}{V_P} = \frac{N_S}{N_P}$$

By appropriate selection of the numbers of turns, a transformer thus allows an alternating voltage to be stepped up—by making N_S more than N_P—or stepped down, by making it less.

Transformers are some of the most efficient electrical "machines", with some large units able to transfer 99.75% of their input power to their output. Transformers come in a range of sizes from a thumbnail-sized coupling transformer hidden inside a stage microphone to huge units weighing hundreds of tons used to interconnect portions of national power grids. All operate with the same basic principles, although the range of designs is wide.

Fig 10.4 Three-Phase, Medium-Voltage Transformer, cut-away view

The conducting material used for the windings depends upon the application, but in all cases the individual turns must be electrically insulated from each other to ensure that the current travels throughout every turn. For small power and signal transformers, in which currents are low and the potential difference between adjacent turns is small, the coils are often wound from enamelled magnet wire. Larger power transformers operating at high voltages may be wound with copper rectangular strip conductors insulated by oil-impregnated paper and blocks of pressboard.

Both the primary and secondary windings on power transformers may have external connections, called taps, to intermediate points on the winding to allow selection of the voltage ratio. The taps may be connected to an automatic on-load tap changer for voltage regulation of distribution circuits.

High temperatures will damage the winding insulation. Small transformers do not generate significant heat and are self-cooled by air circulation and radiation of heat. Power transformers rated up to several hundred kVA can be adequately cooled by natural convective air-cooling, sometimes assisted by fans. Some larger power transformers are immersed in transformer oil that both cools and insulates the windings. The oil is a highly refined mineral oil that remains stable at high temperatures. Liquid-filled transformers to be used indoors must use a non-flammable liquid, or must be located in fire-resistant rooms.

The oil-filled tank often has radiators(Fig 10.4) through which the oil circulates by natural convection; some large transformers employ forced circulation of the oil by electric pumps, aided by external fans or water-cooled heat exchangers. Oil-filled transformers undergo prolonged drying processes to ensure that the transformer is completely free of water vapor before the cooling oil is introduced. This helps prevent electrical breakdown under load.

10.2.2 Power system protection

Good design, maintenance, and proper operating procedures can reduce the probability of occurrence of faults, but cannot eliminate them. Giver that faults will inevitably occur, the objective of protective system design is to minimize their impact.

Faults may have very serious consequences. At the fault point itself, there may be arcing, accompanied by high temperatures and, possibly, fire and explosion. There may be destructive mechanical forces due to very high currents. Overvoltages may stress insulation beyond the breakdown value. Even in the case of less severe faults, high currents in the faulted system may overheat equipment; sustained over heating may reduce the useful life of the equipment.

Clearly, faults must be removed from the system as rapidly as possible. In carrying out this objective, an important, but secondary, objective is to remove no more of the system than absolutely necessary, in order to continue to supply as much of the load as possible. In this connection, we note that temporary loss of lighting or water pumping or air-conditioning load is not usually serious, but loss of service to some industrial loads can have serious consequences. Consider, for example, the problem of repair of an electric arc furnace in which the molten iron has solidified because of loss of power.

Faults are removed from a system by opening or "tripping" circuit breakers (Fig 10.5). These are the same circuit breakers used in normal system operation for connecting or disconnecting generators, lines, and loads. For emergency operation the breakers are tripped automatically when a fault condition is detected. Ideally, the operation is highly selective; only those breakers closest to the fault operate to remove or "clear" the fault. The rest of the system remains intact.

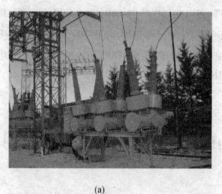
(a)

(b)

Fig 10.5   breaker
(a) circuit breaker, 115 kV; (b) breaker, 3P, 380 V, 20 A

Fault conditions are detected by monitoring voltages and currents at various critical points in the system. Abnormal values individually or in combination cause relays to operate, energizing tripping circuits in the circuit breakers.

10.2.3   Emergency power systems

Emergency power systems(Fig 10.6) are a type of system, which may include lighting, generators and other apparatus, to provide backup resources in a crisis or when regular systems fail. They find uses in a wide variety of settings from residential homes to hospitals, scientific laboratories and modern naval ships. Emergency power systems can rely on generators, batteries, or hydrogen fuel cells. some home emergency power systems use regular lead-acid car batteries, but these do not make a very efficient or reliable system.

With regular generators, an automatic transfer switch is used to connect emergency power. One side is connected to both the normal power feed and the emergency power feed; and the other side is connected to the load designated as emergency. If no electricity comes in on the normal side, the transfer switch uses a solenoid to throw a triple pole, single throw switch. This switches the feed from normal to emergency power. The loss of normal power also triggers a battery operated starter system to start the generator, similar to using a car battery to start an engine. Once the transfer switch is switched and the generator starts, the building's emergency power comes back on(after going off when normal power was lost.)

Fig 10.6　emergency power systems

Unlike emergency lights, emergency lighting is not a type of light fixture; it is a pattern of the building's normal lights that provides a path of lights to allow for safe exit, or lights up service areas such as mechanical rooms and electric rooms. Fire alarm systems and the electric motor pumps for the fire sprinklers is almost always on emergency power. Other equipment on emergency power may include smoke isolation dampers, smoke evacuation fans and outlets in service areas. Hospitals use emergency power outlets to power life support systems and monitoring equipment. Some buildings may even use emergency power as part of normal operations, such as a theater using it to power show equipment because "the show must go on".

10.3　Knowledge about translation（It 的用法）

在英语中，it 是一个十分有用的词，在句子中可以指代各种内容和构成特殊的句型，在分析句子时，往往需要分析 it 所起的作用。

1. it 作代词

（1）it 作无人称代词。it 作无人称代词时可以表示自然现象、天气、时间、距离等，it 是形式上的主语，没有词汇意义，翻译时可省略。

It is fine today.

今天天气很好。

（2）it 作人称代词。it 作人称代词时，用来代替上文中提到的事或物，翻译时可译为它或译成它所代替的事或物。

The zener diode maintains the voltage across its terminals by varying the current that flows through **it**.

稳压管通过改变流过它的电流来维持稳压管二端的电压（稳定）。这里的 it 是指稳压管。

If your microwave oven has an LED or LCD screen and a keypad, it contains a microcontroller.
如果你的微波炉有发光二极管或液晶屏和键盘，它就含有一个单片机。这里的 it 是指微波炉，这种用法与汉语十分相似。

2. it 作形式主语

it 作形式主语时，可代替不定式短语或主语从句，这时 it 称为形式主语。

（1）it 代替不定式的句型有：

it is(was) +形容词+不定式

it +谓语动词+不定式

例：It requires power to drive machines. 开动机器需要能源。It 代替 to drive machines。

（2）it 代替主语从句的句型有：

it is +形容词+主语从句

It is certain that	……是确定的。
It is clear that	……是很清楚的。
It is doubtful that	……是值得怀疑的。
It is desirable that	……是理想的。
It is essential that	……是必要的。
It is possible that	……是可能的。

it +谓语动词的被动态+主语从句

It is well-known that	……是众所周知的。众所周知，……
It is said that	据说……
It is believed that	大家相信……
It is reported that	据报道……
It is generally recognized that	大家公认为
It is supposed that	假设……
It is assumed that	假定……

It has been found that a force is needed to change the motion of a body. 要改变一个物体的运动（状态）需要加外力，这一点已被确定。

It is +名词+主语从句

It is the case that	情况是……
It is a pity that	遗憾的是……
It is no use that	……是没用的。
It is common knowledge that	常识是……

It +不及物动词+主语从句

It now appears that	现在看来……
It seems that	好像是……
It turns out that	显然，……
It happens that	正巧……

3. it 作形式宾语

当动词不定式短语或从句在句中作宾语，而这种宾语又带有补足语时，通常要用 it 放在宾语补足语的前面，使语句简洁明了。

it 代替动词不定式短语

When we want to measure very small currents we find it convenient to use milliamperes and microamperes. 当要测量很小的电流时，我们觉得用毫安和微安是较为方便的。

it 代替宾语从句

The effects we have just discussed make it apparent that here is a means of converting mechanical energy into electrical energy.
根据刚才讨论的效应，显然有一种把机械能转换成电能的方法。

4. it 用于强调句型

强调句型是简单句，可以用来强调句中的主语、宾语和状语，但不能强调谓语和定语。强调语句的句型为：It is(was) +被强调的成分+that (which，who)。在这种句型中，It 和 that 都没有意义，翻译时可在强调成分前加上"正是""就是"等。

It is a pattern of the building's normal lights that provides a path of lights to allow for safe exit.
就是这组大楼的普通指示灯，指示一条通往安全出口的通道。

10.4　Exercises

1. Put the Phrases into English (将下列词组译成英语)
(1) 不可替代的能源
(2) 发电厂
(3) 输电网
(4) 发电机的转子
(5) 机械能
(6) 各种各样的负载
(7) 绝缘问题
(8) 地下电缆
(9) 避雷器
(10) 功率损耗

2. Put the Phrases into Chinese (将下列词组译成中文)
(1) highly flexible in use
(2) water evaporates into steam
(3) the combustion process
(4) protective function
(5) distribution system

(6) overhead transmission

(7) circuit breaker

(8) enhancement of stability

(9) rail system

(10) reduce weight of apparatus

3. Sentence Translation (将下列句子译成中文)

(1) The turbine turns the rotors of the electric generator in whose stator are embedded three (phase) windings.

(2) The voltage is usually "stepped up" by transformers for efficient transmission to remote load centres.

(3) Overhead transmission is used because, for a given voltage level, overhead conductors are much less expensive than underground cables.

(4) The protection equipment includes lightning protectors, circuit breakers and fuses.

(5) One reason for using high transmission-line voltages is to improve energy transmission efficiency.

(6) Near the loads, the transmission voltage is stepped down to the voltages used by loads.

(7) The distribution systems are typically radial because networked systems are more expensive.

(8) The use of lower frequencies also provided the advantage of lower impedance losses, which are proportional to frequency.

4. Translation (翻译) **Surge protector**, 浪涌保护器

Surge protector, as its name implies, the device's primary purpose is to protect electronic devices from the damaging effects of power surges. Also known as transient voltage, surges are any increases above the standard power voltage for a given electrical line. If the increase is large enough, such as those resulting from lightning strikes, the increased voltage can damage the electrical components. After all, if your radio was designed to operate at 120 volts, a 15,000-volt surge will burn through its wiring. Surge protectors handle this problem by blocking or diverting excess current.

10.5 课文参考译文

电能是工业、家庭、商业和交通的最主要的能源之一。电源有许多理想的特点：它是一种清洁的能源（尤其从用电的角度来看），使用方便，很容易从生产地传输到需要用的地方而且适用性广。有时它是一种不可替代的能源。

电能系统由能源，称为发电站（或发电厂），终端用户称为负载和一个连接发电站和负载的输电配电网络组成。最常见的发电站是把燃料或核能或落水的能量转换成电能。

10.5.1 燃料电厂（又称火电厂）

在燃料电厂中，用煤、油或天然气烧锅炉，通过燃烧把锅炉中的水加热变成蒸汽，蒸汽推动汽轮机，汽轮机与发电机有机械上耦合（即汽轮机转动带动发电机转动）。一个典型的燃煤发电厂框图如图10.1所示。

简单地说，电厂的运行是这样的：煤从仓库取出，送入一个粉碎机（或磨粉机），与预先加热的空气混合后喷入锅炉，在锅炉中燃烧。

锅炉中有复杂的管道和鼓状的设备，称为沸腾器，水泵把水压入沸腾器，在沸腾器中水温度上升并蒸发成为蒸汽。蒸汽通过汽轮机，而燃烧后的气体（流动的气体）通过机械和静电除尘器，把超过99%的固体粒子（灰）除去，然后通过烟囱排放出去。

刚才提到的，输入煤粉、空气和水，输出有用的蒸汽的单元有很多个名称，蒸汽发生器、锅炉或沸腾器。当考虑到燃烧的功能时，就用锅炉这个名称，而当考虑水—蒸汽循环的时候则更经常地用沸腾器这个名称。典型的每平方英寸3500磅的压力、1050华氏度的蒸汽通过控制和制动阀送到汽轮机中。控制阀通过调节蒸汽流控制汽轮机—发电机的输出，制动阀起保护作用。

汽轮机带动发电机的转子，发电机的定子中嵌入三相绕组。在这个过程中汽轮机的机械能转换成三相交流电（能），在美国，定子输出有效值为11~30kV，频率为60Hz的线电压。这三相电压一般要通过升压变压器升压，才能高效地通过输电网传到遥远的负载中心。

10.5.2 输配电系统

10.5.1节所描述的电源通常连接到输电系统或输电网，把电能送到各个负载点或负载中心。图10.2为一部分输电网的连接示意图。

发电机（输出的）电压为11~30kV，因为发电机的定子大小有限，要获得更高的输出电压在绝缘方面还有些问题不能解决。然后用变压器升压，升到110~765kV以便于传输，这里的电压都是指线电压（有效值）。到负载端，电压再（通过变压器）降到负载所要求的电压。

负载指的是大负载，如一个城镇或一个大工厂的配电网。这样的配电系统可以提供各种电压级别的电源。大工厂的用户或轨道负载可以直接用23~138kV的高电压电源供电，然后再把电压（根据需要）进一步降压。小工厂或商业用户典型的供电电压为4.16~34.5kV，居民区一般用配电变压器的某一相单相供电，供电电压为120/240V。

输电线路可以是埋在地下或架空线路（图10.3），在城市中常用地下电缆，因为城市里要铺设架空线路成本比较大或者是不可能实现的。地下电缆也常埋在河底下、湖底下和海湾底下。其他地方则用架空线，因为对一个指定的电压级，架空线的成本远低于地下电缆。

配电网把来自输电网的电能传送给用户。配电网一般是点辐射式系统，因为网络式系统成本比较高。配电网上连接的设备包括与输电网相连接的配电变压器、从配电变压器到用户的配电线路、变压器与用户之间的保护控制设备。保护设备包括避雷保护、断路器和熔断器。控制设备包括电压调节器、电容和必要的辅助操作设备。

用高压传输的原因是可以提高能量输送的效率。因为所传输的有功功率等于电流、电压和功率因数（相位差ϕ的余弦）的积，通过增加电压，减少电流可以传输相同的功率。为什么要减少传输电流呢？因为流过导体的有功功率损耗是导体电阻和电流平方的乘积，用公式

$P=I^2R$ 表示,这就意味着在给定的导线上传输一定的有功功率时,如果电流加倍,则功率损耗是原来的四倍。所以在用高电压在电网上传输大量的功率是很有好处的。

用高压传输的另一个原因是增加传输网的稳定性。

10.5.3 交流电源的频率

不同国家的电网频率是不同的,大部分国家电网频率是 50~60Hz,有些国家如日本比较特别,有 50Hz 和 60Hz 两种频率的电网。

电网采用低频的好处是使得低速电动机的设计变得比较容易,因为在起重机、压碾机和辊压机,以及轨道交通中所用的换向器式的牵引电动机都需要用低速电机,但低频电网的缺点是造成白炽灯和荧光灯产生能察觉到的闪烁现象。在一些欧洲的轨道系统(如澳大利亚、德国、挪威、瑞典和瑞士)中目前还在用 16.6Hz 的电源。电网采用较低频率还有这样的好处:阻抗损耗比较低,因为阻抗(电抗)是正比于频率的。加拿大的尼亚加拉水电站建设初期考虑到低频对牵引电动机和大型感应电动机有利,设计成电源的频率为 25Hz,并提供给白炽灯用以照明(虽然有明显的闪烁现象)。到 20 世纪 50 年代大部分尼亚加拉居民用户和商业用户的电源频率从 25Hz 转换成 60Hz,虽然工业用户直到 21 世纪初仍用 25 Hz 的电源。

军工、纺织工业、矿业、大型计算机、飞机、航天飞机有时用 400Hz 的电源,其好处是可以减轻设备的重量或提高电动机的转速。

10.6 阅读材料参考译文

10.6.1 变压器

变压器(图 10.4)是把一个电路的电能通过(电磁)感应耦合的方式转换到另一个电路的设备。在第一个线路(一次绕组)中的交变电流产生了一个交变的电磁场,这个电磁场反过来在第二个电路(二次绕组)中感应产生交变的电压。当在二次绕组端加上负载时,就有交变电流从变压器流出,把一个电路中的电能传递到了另一个电路。

一个理想变压器的次绕组的感应电压 V_S 的大小与原绕组的电压 V_P 成比例,这比例与这两个绕组相应的线圈匝数之比相等。

$$\frac{V_S}{V_P} = \frac{N_S}{N_P}$$

通过选择绕组匝数,使次绕组匝数 N_S 大于原绕组匝数 N_P,构成升压变压器,如果二次绕组匝数 N_S 小于原绕组匝数 N_P,就构成降压变压器。

(电力)变压器是效率最高的电气设备,大型的电力变压器的可把输入功率的 99.75% 转换成输出功率。变压器的种类很多,小到可放在一个微型话筒中的只有指甲大小的耦合变压器,到国家电网中所连接的几百吨重的巨型变压器。虽然变压器的种类很多,但基本工作原理是相同的。

绕组所需用的导体材料根据应用场合决定，但每一圈绕组之间必须是电气绝缘的，以保证电流流过每个绕组。对小功率信号变压器来说，电流很小，相邻线圈之间的电势差（电压）是很小的，线圈通常是用漆包线绕成的。较大功率的变压器的工作电压比较高，线圈是用外包油浸纸和纸板块进行绝缘隔离的扁铜条绕成的。

电力变压器的一次绕组和二次绕组的绕组上可能带有外部抽头，称为接线端，可以在变压器外部连接以选择不同的电压比。接线端可以连接到一个自动随负载变化的调节设备上，用来稳定配电线路中的电压。

高温会损坏绕组的绝缘性能，小变压器产生的热量并不大，可以通过空气流动和热辐射的形式自我冷却。额定功率高达几百 kVA 电力变压器也可以通过空气冷却方式得到充分冷却，有时要加风扇帮助冷却。更大的电力变压器是（把绕组）浸在变压器（绝缘）油中，对绕组起到冷却和绝缘两重作用。变压器油是高度提纯的矿物油可以在高温下保持稳定性。用于室内的浸在液体中的变压器要用一种不易燃的液体，或者必须放在防火房间内。

油箱一般有散热器（图 10.4），油通过散热器作自然循环，有些大的变压器还用电泵强迫油循环，加上外部的风扇或冷却水热交换器。为了保证变压器中完全没有水蒸气，油浸变压器在冷却油流入之前要经过相当长的干燥过程。这样可以避免接上负载后出现电击穿（短路）。

10.6.2 电力系统保护

好的设计、维护和适当的操作可以减少故障出现的可能性，但并不能完全消除故障。既然故障是不可避免的，电力系统保护设计的目标就是使故障危害最小化（尽可能减少故障的危害）。

故障可能有非常严重的结果，在故障点，可能出现电弧，并伴随高温，可能引起火灾甚至爆炸。由于大电流还可能出现毁灭性的机械力。过电压可能使得绝缘被击穿。即使出现不太严重的故障时，在故障系统中的大电流也会使设备过热，设备长期过热会减少设备的使用寿命。

显然，系统中的故障部分必须尽快切除。在实现这个目标的同时，一个仅次于它的也很重要的目标是为了继续对尽可能多的负载供电，除了绝对有必要需要断开的电路，不要过多地切断系统。在电网的连接中，注意到暂时断开照明、水泵或空调负载可能并没有多大关系，但有些工业负载如果突然断电，会造成严重后果。试想一下，因为断电使得熔化的铁熔液在电弧炉中结成固体，这种情况是很难修复的。

故障是通过断开或"触发"电路断路器的方法从系统切除的（图 10.5）。在正常电网系统操作中也是用同样的电路断路器接上或断开发电机、电线和负载的。当故障出现时的紧急操作中，断路器是自动动作的。在理想情况下，做什么操作是经过选择的，只有那些最靠近故障点的断路器动作，切除或"清除"故障。系统的其他部分仍保持工作。

通过监测电力系统中各个关键点的电压和电流可以发现故障，一个或几个不正常的测量值会使继电器动作，使电路断路器的触发电路得电（使断路器动作）。

10.6.3 应急电源

应急电源（图 10.6）是指一个包括照明、发电机和其他设备的系统，用来作为正常电力系统出故障或关键时刻的后备电源。从居民区到医院、科学实验室和现代海军军舰上都会用到应急电源。应急电源可以用发电机、电池或氢燃料电池。有些家用应急电源是用普通的铅酸车用电池，但其效率和可靠性都不高。

　　一个自动的切换开关用于把普通的发电机与应急系统相连接，一边是连接普通电网输入端和应急电网输入端，另一端是与作为应急时要用的负载相连接。如果在普通电网端没有输入时，切换开关利用螺线管合上一个三极性的单刀开关，从电网输入切换到应急电源输入。普通电网的失电还触发电池启动系统，使发电机开始工作，类似于用汽车的电池启动发动机。一旦切换开关切换动作且发电机开始工作，大楼的应急电源开始恢复供电（在普通电网失电时断电后）。

　　与应急灯不同，应急照明不是一种照明设施（它只是与应急电源相连的指示灯），就是这组大楼的普通指示灯，指示一条通往安全出口的通道，或指示某些服务设施处如机械室、电气室。消防警报系统和用于消防洒水的抽水泵电机也是连在应急电源上的，其他连接应急电源的设备还有烟隔离装置、烟排空风扇和服务区的电源插座。医院用应急电源作为生命保障系统和监视设备的电源，有些建筑中甚至可能用应急电源作为正常照明电源的一部分，如剧院用它作为放映设备的电源，因为"放映不能中断"。

Unit 11 Microcontroller

11.1 Text

Microcontrollers[1] are hidden inside a surprising number of products these days (Fig 11.1). About 55% of all CPUs sold in the world are 8-bit microcontrollers. If your microwave oven has an LED or LCD screen and a keypad, it contains a microcontroller. All modern automobiles contain at least one microcontroller, and can have as many as six or seven: The engine is controlled by a microcontroller, as are the anti-lock brakes and so on. Any device that has a remote control almost certainly contains a microcontroller: TVs, VCRs and high-end stereo systems all fall into this category. Digital cameras, cell phones, camcorders, answering machines, laser printers, telephones (the ones with caller ID, 20-number memory, etc), and refrigerators, dishwashers, washers and dryers (the ones with displays and keypads).....You get the idea. Basically, any product or device that interacts with its user has a microcontroller buried inside.

Fig 11.1   products contained a microcontroller

11.1.1   What is Microcontroller?

A microcontroller (also MCU or μC) is a computer-on-a-chip, containing a processor, memory, and input/output functions. It is a microprocessor emphasizing high integration, in contrast to a general-purpose microprocessor (the kind used in a PC). In addition to the usual arithmetic and logic elements of a general purpose microprocessor, the microcontroller integrates additional elements such as read-write memory for data storage, read-only memory for program storage, EEPROM for permanent data storage, peripheral devices, and input/output interfaces. At clock speeds of as little as a few MHz to tens MHz, microcontrollers often operate at very low speed compared to modern day microprocessors, but this is adequate for typical applications. They consume relatively little power (milliwatts), and will generally have the ability to sleep while waiting for an interesting peripheral event such as a button press to wake them up again to do something. Power consumption while sleeping may be just nanowatts, making them ideal for low power and long lasting battery applications.

11.1.2　Intel 8051

The Intel 8051 is a Harvard architecture[2], single chip microcontroller (μC) which was developed by Intel in 1980 for use in embedded systems.

Intel's original 8051 family was developed using NMOS technology, but later versions, identified by a letter "C" in their name, e.g. 80C51, used CMOS technology and were less power-hungry than their NMOS predecessors-this made them eminently more suitable for battery-powered devices.

The 8051 families have following important features(Fig 11.2):

(1) 8-bit data bus—It can access 8 bits of data in one operation (hence it is an 8-bit microcontroller)

(2) 16-bit address bus—It can access 2^{16} memory locations—64 kB each of RAM and ROM

(3) On-chip RAM—128 bytes ("Data Memory")

(4) On-chip ROM—4 kB ("Program Memory")

(5) Four byte bi-directional input/output port

(6) UART (serial port)

(7) Two 16-bit Counter/timers

(8) Two-level interrupt priority

(9) Power saving mode

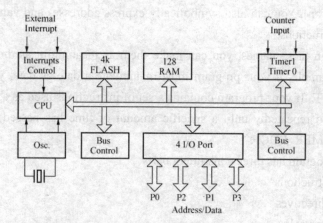

Fig 11.2　important features in 8051 families

A particularly useful feature of the 8051 core is the inclusion of a boolean processing engine which allows bit-level boolean logic operations to be carried out directly and efficiently on internal registers and RAM. This feature helped to cement the 8051's popularity in industrial control applications. Another valued feature is that it has four separate register sets, which can be used to greatly reduce interrupt latency compared to the more common method of storing interrupt context on a stack.

The original 8051 core ran at 12 clock cycles per machine cycle, with most instructions executing in one or two machine cycles. With a 12 MHz clock frequency[3], the 8051 could thus execute 1 million one-cycle instructions per second or 500 000 two-cycle instructions per second. Enhanced 8051 cores are now commonly used which run at six, four, two, or even one clock per machine cycle, and have clock frequencies of up to 100 MHz, and are thus capable of an even greater number of instructions per second.

Common features included in modern 8051 based microcontrollers include built-in reset timers with brown-out detection, on-chip oscillators, self-programmable Flash ROM program memory, boot loader code in ROM, EEPROM non-volatile data storage, I^2C, SPI, and USB host interfaces, PWM generators, analog comparators, A/D and D/A converters, extra counters and timers, in-circuit debugging facilities, more interrupt sources, and extra power saving modes.

11.1.3 What is an Assembler?

An assembler is a software tool designed to simplify the task of writing computer programs. It translates symbolic code into executable object code. This object code may then be programmed into a microcontroller and executed. Assembly language programs translate directly into CPU instructions which instruct the processor what operations to perform. Therefore, to effectively write assembly programs, you should be familiar with both the microcomputer architecture and the assembly language.

Assembly language operation codes are easily remembered (MOV for move instructions, ADD for addition, and so on). You can also symbolically express addresses and values referenced in the operand field of instructions.

Since you assign these names, you can make them as meaningful as the mnemonics for the instructions. For example, if your program must manipulate a date as data, you can assign it the symbolic name DATE. If your program contains a set of instructions used as a timing loop (a set of instructions executed repeatedly until a specific amount of time has passed), you can name the instruction group TIMER_LOOP.

An assembly program has three constituent parts:

(1) Machine instructions

(2) Assembler directives

(3) Assembler controls

A machine instruction is a machine code that can be executed by the machine. Detailed discussion of the machine instructions can be found in the hardware manuals of the 8051 family or other microcontrollers.

Assembler directives are used to define the program structure and symbols, and generate non-executable code (data, messages, etc).

Assembler controls set the assembly modes and direct the assembly flow.

Review

(1) Basically, any product or device that interacts with its user has a microcontroller buried inside.
(2) A microcontroller is a computer-on-a-chip, containing a processor, memory, and input/output functions.
(3) The Intel 8051 families have many important features.
(4) An assembler is a software tool designed to simplify the task of writing computer programs.

Notes to the text

[1] Microcontroller 从字面译为微控制器，中文常译成单片机，有时与microprocessor不加区别，译成微处理器。
[2] Harvard architecture 哈佛结构，是计算机的一种结构形式。
[3] clock frequency 时钟频率 f 与时钟周期 T 的关系是：$T=1/f$。

Technical Words

arithmetic [əˈrɪθmətɪk] n. 算术，算法
assembly [əˈsembli] n. 集合，装配，集会，集结，汇编，这里指汇编
consumption [kənˈsʌm(p)ʃ(ə)n] n. 消费，消费量，肺病
debug [diːˈbʌɡ] vt. 调试、跟踪、并排除计算机软件中的错误
eminently [ˈemɪnəntli] adv. 不寻常地
enhance [ɪnˈhæns] v. 增强（提高，放大），增进，增加
latency [ˈleɪtənsi] n. 锁存器，潜伏，潜在，潜伏器
microcontroller [ˌmaɪkrəʊkənˈtrəʊlə] n. 微控制器，单片机
mnemonic [nɪˈmɒnɪks] adj. 记忆的，记忆术的 n. 记忆码
nanowatt [ˈnænəˌwɒt] nano+watt 毫微瓦（特），10^{-9} 瓦特
peripheral [pəˈrɪf(ə)r(ə)l] adj. 外围的 n. 外围设备

Technical Phrases

The 8051 families	8051 系列
8-bit microcontroller	8 位单片机
anti-lock brake	防抱死装置（汽车刹车技术）
answering machines	应答机

general-purpose microprocessor	通用型微处理器
embedded systems	嵌入式系统
Flash ROM program memory	闪存程序存储器
assembly language	汇编语言
brown-out detection	布朗检测器，低电压检测器，掉电检测器

11.2　Reading materials

11.2.1　Microcontroller VS computer

A microcontroller is a computer. All computers—whether we are talking about a personal desktop computer or a large mainframe computer or a microcontroller—have several things in common:

All computers have a CPU (central processing unit) that executes programs. If you are sitting at a desktop computer right now reading an article, the CPU in that machine is executing a program that implements the Web browser that is displaying the article.

The CPU loads the program from somewhere. On your desktop machine, the browser program is loaded from the hard disk.

The computer has some RAM (random-access memory) where it can store "variables."

And the computer has some input and output devices so it can talk to people. On your desktop machine, the keyboard and mouse are input devices and the monitor and printer are output devices. A hard disk is an I/O device—it handles both input and output.

The desktop computer you are using is a "general purpose computer" that can run any of thousands of programs. Microcontrollers are "special purpose computers". Microcontrollers do one thing well. There are a number of other common characteristics that define microcontrollers. If a computer matches a majority of these characteristics, then you can call it a "microcontroller":

Microcontrollers are "embedded" inside some other device (often a consumer product) so that they can control the features or actions of the product. Another name for a microcontroller, therefore, is "embedded controller".

Microcontrollers are dedicated to one task and run one specific program. The program is stored in ROM (read-only memory) and generally does not change.

Microcontrollers are often low-power devices. A desktop computer is almost always plugged into a wall socket and might consume 50 watts of electricity. A battery-operated microcontroller might consume 50 milliwatts.

A microcontroller has a dedicated input device and often (but not always) has a small LED or LCD display for output. A microcontroller also takes input from the device it is controlling and controls the device by sending signals to different components in the device.

For example, the microcontroller inside a TV takes input from the remote control and displays output on the TV screen. The microcontroller controls the channel selector, the speaker system and certain adjustments on the picture tube electronics such as tint and brightness. The engine controller in a car takes input from sensors such as the oxygen and knock sensors and controls things like fuel mix and spark plug timing. A microwave oven controller takes input from a keypad, displays output on an LCD display and controls a relay that turns the microwave generator on and off.

A microcontroller is often small and low cost. The components are chosen to minimize size and to be as inexpensive as possible.

A microcontroller is often, but not always, ruggedized in some way. The microcontroller controlling a car's engine, for example, has to work in temperature extremes that a normal computer generally cannot handle. A car's microcontroller in Alaska has to work fine in −30 degree °F (−34 °C) weather, while the same microcontroller in Nevada might be operating at 120 degrees °F (49 °C). When you add the heat naturally generated by the engine, the temperature can go as high as 150 or 180 degrees °F (65~80 °C) in the engine compartment.

On the other hand, a microcontroller embedded inside a VCR hasn't been ruggedized at all.

11.2.2 Modular Programming

Many programs are too long or complex to write as a single unit. Programming becomes much simpler when the code is divided into small functional units. Modular programs are usually easier to code, debug, and change than monolithic programs.

The modular approach to programming is similar to the design of hardware that contains numerous circuits. The device or program is logically divided into "black boxes" with specific inputs and outputs. Once the interfaces between the units have been defined, the detailed design of each unit can proceed separately.

The benefits of modular programming are:

Efficient Program Development: programs can be developed more quickly with the modular approach since small subprograms are easier to understand, design, and test than large programs. With the module inputs and outputs defined, the programmer can supply the needed input and verify the correctness of the module by examining the output. The separate modules are then linked and located by the linker into an absolute executable single program module. Finally, the complete module is tested.

Multiple Use of Subprograms: code written for one program is often useful in others. Modular programming allows these sections to be saved for future use. Because the code is relocatable, saved modules can be linked to any program which fulfills their input and output requirements. With monolithic programming, such sections of code are buried inside the program and are not so available for use by other programs.

Ease of Debugging and Modifying: modular programs are generally easier to debug than monolithic programs. Because of the well defined module interfaces of the program, problems can be isolated to specific modules. Once the faulty module has been identified, fixing the problem is considerably simpler. When a program must be modified, modular programming simplifies the job. You can link new or debugged modules to an existing program with the confidence that the rest of the program will not change.

11.2.3 Multi-Core processors

In October 1989, envisioning the future through the lens of Moore's Law, four Intel technologists authored an article entitled "Microprocessors Circa 2000" which predicted that multi-core processors could come to market soon after the turn of the century. Fifteen years later their predictions proved true, and multi-core processor capability development had become one of the top business and product initiatives for both Intel and rival chipmaker AMD.

Explained most simply, multi-core processor architecture entails silicon design engineers placing two or more processor-based computational engines within a single processor. This multi-core processor plugs directly into a single processor socket, but the operating system perceives each of its execution cores as a discrete logical processor with all the associated execution resources.

By spreading the computational work performed by a single microprocessor core in traditional microprocessors over multiple execution cores, a multi-core processor can perform more work within a given clock cycle. To realize this performance gain, the software running on the platform must be written such that it can spread its workload across multiple execution cores. This functionality is called thread-level parallelism or "threading". Applications and operating systems that are written to support it are referred to as "threaded" or "multi-threaded".

A processor equipped with thread-level parallelism can execute completely separate threads of code. This can mean one thread running from an application and a second thread running from an operating system, or parallel threads running from within a single application. Thread-level parallelism is of particular benefit to many multimedia applications because many of their operations are capable of running in parallel.

When combined with Hyper-Threading, Intel dual-core processors will be able to process four software threads simultaneously(Fig 11.3) by more efficiently using resources that otherwise may have remained idle.

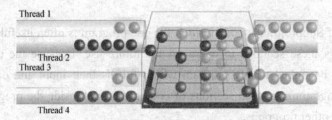

Fig 11.3　to process four software threads simultaneously

11.3 Knowledge about translation（That 的用法）

That 的用法很多，在英语阅读和翻译中，正确区分 that 的作用，可以更好地理解一些复杂的句子。下面对 that 的主要用法作一介绍。

1. That 作指示代词

That 用作指示代词时，作定语相当于一个形容词修饰它后面的名词，作主语、表语和宾语时相当于一个名词，在科技文章中常用它来代替句中已出现过的某一名词，以免重复。

The proton has a single positive charge equal to that of an electron which is negative.

质子带有一个正电荷，其电量与电子所带的负电荷相等。（that 指代前面的 charge，which is negative 为定语，修饰 that 即 charge）。

That 也可以用来指代整个句子：

Electrical energy can be changed easily and directly into all the other forms of energy. That is why electricity is so useful in everyday life.

电能可以方便而直接地转换成其他形式的能量，这就是电在日常生活中如此有用的原因。（that 指前面整个句子。）

2. That 作关系代词引出定语从句

That 用作关系代词引出定语从句与 that 引出其他从句在语法结构上不同，that 在引出的定语从句中是一个语法成分，作主语或宾语。

Electrical engineering, sometimes referred to as electrical and electronic engineering, is a field of engineering that deals with the study and application of electricity, electronics and electromagnetism.

电气工程有时称为电气和电子工程，是研究电、电子和电磁效应及其应用的一个工程领域。（其中 that 在从句中为主语，指电气工程。）

3. That 作为连词引出名词性从句

这时 that 本身无词汇意义，也无复数形式，在句中只起语法上的连接作用。

It is becoming more common that an application-specific IC (ASIC) is required, at least for high volume applications.

专用集成电路的需要已更加普遍，至少对大批量的应用器件来说更需要专用的集成电路。这里因主语很长，用 it 作形式主语，用 that 引出主语从句。

A well-known characteristic of light is that it travels in straight lines. 光的一个众所周知的特征是它以直线传播。

有一些名词常用 that 引出的同位语从句来加以说明，这样的名词有：fact（事实），idea（思想，概念），theory（理论）conclusion（结论），discover（发现）等。

From this we come to the conclusion that the resistance depends on the sort of material of which the electric conductor is made.

由此我们得出结论，电阻取决于导体是由哪种材料做成的。

The importance of this concept lies in the fact that most of the generation and distribution of electric power is accomplished with polyphase systems.

这个概念的重要性在于多数发电机和配电网都是用多相系统实现的。（the fact that，为便于理解，也可以译成：因为多数发电机和配电网都是用多相系统实现的，因此这个概念很重要。）

4. In that 的用法

Liquids are different from solids in that liquids have no definite shape.

液体与固体的区别在于，液体没有固定的形状。（注意一般在介词后面是不能用 that 的，但 in that 是个特例，可以记成固定用法，译为：因为……，表示一些内在的原因。）

5. 连词引出状语从句

常用的结构形式有：so that……（导致，结果是），so +形容词（或副词）+that（如此……以至于），such that ……或 such +名词+that（这样……以至于，）引出结果状语从句。

Some atoms are so constructed that they lose electrons easily.

有些原子的结构使它很容易失去电子。（如此构建以至于……）

When the water level drops enough so that the tank empty switch is on (up), the PLC will open the valve to let more water in.

当水位掉到使得水箱空开关动作（动合触点闭合）时，PLC 就打开阀门让水流入。

Without such an increase in productivity that the design of complex systems might not be achievable within a reasonable time frame.

如果在（芯片设计）生产率方面没有这样一个增加，复杂系统的设计将不可能在一个合理的时间阶段内实现。

11.4 Exercises

1. Put the Phrases into English (将下列词组译成英语)
(1) 激光打印机
(2) 8 位单片机
(3) 数据存储
(4) 功率消耗
(5) 独立寄存器组
(6) 机器周期
(7) 模拟比较器
(8) 中断源
(9) 汇编语言
(10) 操作代码

2. Put the Phrases into Chinese (将下列词组译成中文)

(1) microwave oven

(2) remote control

(3) digital cameras

(4) interacts with its user

(5) general-purpose microprocessor

(6) permanent data storage

(7) battery-powered device

(8) Four byte bi-directional input/output port

(9) internal registers

(10) clock frequencies

3. Sentence Translation (将下列句子译成中文)

(1) It is a microprocessor emphasizing high integration, in contrast to a general-purpose microprocessor (the kind used in a PC).

(2) They consume relatively little power (milliwatts), and will generally have the ability to sleep function.

(3) This made them eminently more suitable for battery-powered devices.

(4) A particularly useful feature of the 8051 core is the inclusion of a boolean processing engine which allows bit-level boolean logic operations to be carried out directly.

(5) Assembly language programs translate directly into CPU instructions which instruct the processor what operations to perform.

4. Translation (翻译)

Server is a computer or device on a network that manages network resources. For example, a file server is a computer and storage device dedicated to storing files. Any user on the network can store files on the server. A print server is a computer that manages one or more printers, and a network server is a computer that manages network traffic. A database server is a computer system that processes database queries.

Servers are often dedicated, meaning that they perform no other tasks besides their server tasks. On multiprocessing operating systems, however, a single computer can execute several programs at once. A server in this case could refer to the program that is managing resources rather than the entire computer.

11.5　课文参考译文

近年来在数不清的产品内部都有单片机（图 11.1），世界上所销售的 CPU 中大约有 55% 是 8 位的单片机。如果你的微波炉有发光二极管或液晶屏和键盘，它就含有一个单片机。当代所有的汽车都至少有一个单片机，有的多达 6 或 7 个单片机：发动机是用单片机控制的，和防抱死制动（刹车）系统也是用单片机控制的，还有其他地方也要用单片机控制。几乎所有你能想到的可以遥控的设备一定都含有一个单片机：电视机、录放像机、高端立体声系统都属于这一类。数码相机、手机、摄像机、自动应答设备、激光打印机、电话（含有来电显示、号码储存等功能）、冰箱、洗碗机、洗衣机和干衣机（带有显示和键盘的）……基本上所有能与用户交互的设备都内嵌单片机。

11.5.1 单片机是什么？

微控制器（也称 MCU 或 μC）是单片机，它含有处理器、存储器和输入/输出功能。与通用的微处理器（在个人计算机中所用的微处理器）比较，它是高度集成的微处理器。除了通用微处理器所含有的运算和逻辑单元外，单片机还集成了其他单元，如可储存数据的读写存储器，可储存程序的只读存储器，可永久储存数据的 EEPROM，外围设备和输入/输出接口。其时钟速度只有几 MHz 到几十 MHz，与现在的微处理器相比，单片机的运行速度很慢，但对一些单片机典型的应用来说是足够了。单片机消耗功率很小（毫瓦级）且一般有睡眠功能，即（不用时处于睡眠）等待一个外部事件如按键按下唤醒单片机继续工作。睡眠时消耗功率只有纳瓦级（10^{-9}W），所以很适宜用在低功耗且长期用电池供电的场合。

11.5.2 Intel 8051

英特尔 8051 是哈佛结构的单片机，是英特尔公司在 1980 年开发的用于嵌入式系统的芯片。

英特尔最早的 8051 系列是用 NMOS 技术开发的，但后来的标以 "C" 字母如 80C51 是用 CMOS 技术开发的芯片，它比以前的 NMOS 芯片更省电，更适用于电池供电系统。

8051 系列有以下重要特征（图 11.2）：

（1）8 位数据总线——可一次访问 8 位数据（因此称为 8 位单片机）。
（2）16 位地址总线——可以访问 2^{16} 个地址，分别是 64k RAM 和 64 k ROM。
（3）片内 RAM——128 位（"数据存储"）。
（4）片内 ROM——4kB（"程序存储"）。
（5）四个（8 位）双向输入/输出端口。
（6）UART（串行口）。
（7）两个 16 位计数/定时器。
（8）两级带优先权的中断。
（9）省电模式。

8051 内核的一个特别有用的特点是含有一个逻辑处理部分，它可以对内部寄存器和 RAM 直接进行位逻辑操作，这个特点使 8051 在工业控制应用中十分有用。另一个有价值的特点是它有四个分立的寄存器组（$R_0 \sim R_3$），与一般把中断的内容存储在堆栈的方法比较，可以减小中断响应时间。

最早的 8051 内核运行的每个机器周期含 12 个时钟周期，大部分指令的执行需要 1~2 个机器周期， 因此，当 8051 的时钟频率为 12MHz 时，每秒可执行 1 百万条 1-机器周期的指令，或 50 万条 2-机器周期指令。现在增强型 8051 的内核有每机器周期只含 6 个、4 个、2

个甚至 1 个时钟周期，时钟频率甚至高达 100MHz，因此每秒可以执行更多的指令。

现在基于 8051 系列的单片机芯片的共同特点包括如下：内部含有掉电检测功能的复位定时器，芯片内振荡器，可直接编程的闪存 ROM 程序存储器，在 ROM 中有引导程序，EEPROM 稳定的数据存储，I^2C、SPI 和 USB 主机接口界面，PWM（脉宽调节）发生器，模拟比较器，模/数和数/模转换器，额外的计数器和定时器，在线调试工具，更多的中断源和特别的省电模式。

11.5.3 汇编程序是什么？

汇编程序是一种软件工具，用来简化编写计算机程序的，汇编程序把符号化的代码（汇编指令）编译成可执行的目标代码（机器码）。然后把目标代码输入到单片机（的程序存储器）中并加以执行。汇编语言程序可直接译成 CPU 指令，命令微处理器完成相应的运算。所以要有效地编写汇编语言程序，就需要对微型计算机的结构和汇编语言都很熟悉。

汇编语言操作指令很容易记（MOV 就是移动指令，ADD 就是加法指令等）。还可以用符号表示（存储器）地址和指令的操作字段值。

因为指定了这些名字，指令就有了意义，比较容易记忆。如果你在程序中要多次用到一个日期作为数据，你可以给这个日期指定一个符号化的名字 DATE（日期）。如果你程序中含有一组指令作为一个延时循环（一组指令要反复执行，直到经过指定的时间，起到延时的作用），你可以给这段指令起名 TIMER-LOOP（定时循环）。

汇编程序有机器指令、汇编指令、汇编控制三个要素。

机器指令是机器代码，可以由 CPU 执行的。机器指令的详细讨论可以参看 8051 系列或其他单片机的硬件手册。

汇编指令用于定义程序的结构和符号，产生非执行代码（数据，信息等）。

汇编控制置设置汇编的模式和汇编程序的流向。

11.6 阅读材料参考译文

11.6.1 单片机与计算机

单片机就是一个计算机，无论讨论的是个人台式计算机、大型计算机或一个单片机，所有的计算机都有一些共同点：

所有的计算机都有一个 CPU（中央处理器）用于执行程序。如果你现在正坐在台式计算机前读一篇文章，台式机中 CPU 就在执行一个程序用于打开显示这篇文章的网络浏览器。

CPU 要从某个存储器中下载它所要执行的程序，在你的台式机中，浏览器程序就是从硬盘中下载的。

计算机都有一些 RAM（随机存储器），可以用来存放"变量"。

计算机还有一些输入和输出设备，可以和人们交流信息。在台式机中，键盘和鼠标就是输入设备，显示器和打印机是输出设备。硬盘也是一种输入/输出设备——可以输入和输出（信息）。

台式机是一种通用型的计算机，可用来执行很多程序。单片机是一种有指定目标的计算机，单片机只做一件事。还有很多其他定义单片机的其他共同特点，如果一台计算机符合以下这些特点中的大部分，就可以称作是一台单片机：

单片机是嵌入在某种其他设备中的（通常是消费产品），所以单片机可以控制该产器的动作或性能。单片机也就被称作为嵌入式控制器。

单片机指定完成某个任务，执行某个指定的程序。这程序储存在 ROM（只读存储器）中通常（工作时程序）不再改变。

单片机一般是低功耗器件，大部分台式机总要用交流电源供电，功率消耗可能达 50W，而靠电池供电的单片机可能只消耗 50mW 功率。

单片机有指定的输入设备，通常（但也有例外）有小的发光二极管或液晶显示器用于输出。单片机还可以从它所控制的设备中得到输入信号并通过输出信号给设备的不同部件来控制该设备。

例如：电视机中的单片机从遥控器中获得输入信息，并在电视机屏幕上显示输出（信息）。单片机控制选择频道，音响系统和调节显像管上画面的显示，如色彩和亮度。汽车里的发动机控制器（单片机）从传感器（如氧气传感器和振动传感器）中接收输入信号，从而控制燃料的混合，点火电脉冲的时间宽度等。微波炉中的单片机接收来自键盘的输入信号，在液晶显示器显示输出及并控制微波发生器的通电和切断时间。

单片机一般比较小，且价格便宜。器件一般是选取微小化，且尽可能便宜。

单片机一般（但也有例外）比较坚固耐用。例如，控制一辆汽车的发动机的单片机要在普通计算机所不能承受的环境温度中工作，阿拉斯加州（美国州名）中汽车中的单片机能在零下 30°F（−30℃）的环境中工作，而同样的单片机也可以在内华达州（美国西部内陆州）的 120°F（49℃）的高温下工作。再加上发动机产生的热量，发动机中的温度可能高达 150～80°F（65～80℃）。

在另一方面，嵌入录像机中的单片机却一点也不耐用。

11.6.2　模块化编程

很多程序太长或太复杂，不能作为一个程序编写。如果把程序代码分成小的功能模块，编程就变得比较简单了。在编程、调试和修改时，模块化编程一般比编写单一程序容易。

模块化方法编程有点类似于含有多个电路的硬件设计。器件或程序在逻辑上分成指定输入和输出的"黑箱"。一旦定义了单元之间的接口关系，每个单元的详细设计可以分别进行。

模块化编程的好处如下：

开发程序效率高：用模块化的方法开发程序比较快，因为小的子程序比大的程序容易理解、设计和测试。定义了模块的输入和输出后，程序员可以给出程序所需要的输入，并通过检查输出来验证模块的正确性。然后把分立的模块用连接（程序）确定各模块的存放位置，并连接起来成为一个可执行的单一程序。最后，完整的模块再进行测试。

可多次调用子程序：一个程序中所编写的子程序往往在其他程序中也有用。模块化编程可以把这些程序段（子程序）保存下来，以后再调用。因为各模块（子程序）中的代码是可以重新定位的，保存的模块（子程序）可以被任何满足它们的输入和输出要求的程序调用。对单一程序来说，这样的程序段是含在程序的内部，不能这样被其他程序调用。

调试和修改比较容易：模块化编程一般比单一程序容易调试。因为程序很好地定义了模块的输入输出接口，出现的问题时可以隔离相应的模块。一旦有错的模块被确定，要修改程序就比较简单（只需要调试这个模块）。当一个程序（的功能）需要修改时，模块化编程可以比较容易处理。可以把一个已有的程序和一个新的或已调试过的模块相连接，程序的其他部分都不需要改变。

11.6.3 多核处理器

在1989年十月，根据摩尔定律，四个英特尔的技术专家发表了一篇题为"约在2000年诞生的处理器"的文章，预言在世纪交变之际，多核处理器将很快在市场上出现。十五年后他们的预言成真，对英特尔和与其竞争的芯片制造商AMD，多核处理器性能开发已成为他们主要业务之一，并有产品面市。

最简单地说，多核处理器结构要求硅芯片设计工程师把两个或更多个基于处理器的计算内核放在一个处理器中。这个多核的处理器直接插入一个单一处理器的插座，但操作系统能把它的每个执行内核作为一个与所有指定执行的资源相连接的独立的逻辑处理器。

通过把传统微处理器中一个内核完成的计算工作分摊给多个内核执行，一个多核处理器可以在一个给定的时钟周期中执行更多的任务。为了实现这个性能，在这个平台上所运行的软件要编写成具有把任务分散给多个内核执行的功能的软件，这种功能称为平行"线程"。支持这种功能的操作系统和应用软件就称为线程或多线程的。

带有多线程的处理器可以执行完全分开的代码线程，这就意味着为一个应用程序运行的一个线程和为一个操作系统运行的第二个线程或在一个应用程序中两个平行线程同时运行。对多媒体应用程序来说多线程运行有特别的好处，因为他们的很多运算是可以平行运行的。

结合"超线程"，为了更有效地利用资源，英特尔的双核处理器需要可以同时处理四个软件线程（图11.3），否则可能造成资源的闲置（没有充分利用）。

Unit 12   Programmable Controller

12.1   Text

A programmable logic controller, PLC, or programmable controller is a digital computer used for automation of typically industrial electromechanical processes, such as control of machinery on factory assembly lines, amusement rides, or light fixtures. PLCs are designed for multiple arrangements of digital and analog inputs and outputs, extended temperature ranges, immunity to electrical noise, and resistance to vibration and impact. A PLC is an example of a "hard" real-time system since output results must be produced in response to input conditions within a limited time, otherwise unintended operation will result.

12.1.1   PLC Introduction

At the outset of industrial revolution, especially during sixties and seventies, relays were used to operate automated machines, and these were interconnected using wires inside the control panel. In some cases a control panel covered an entire wall. To discover an error in the system much time was needed especially with more complex process control systems. On top of everything, a lifetime of relay contacts was limited, so some relays had to be replaced. If replacement was required, machine had to be stopped and production too. Also, it could happen that there was not enough room for necessary changes. Control panel was used only for one particular process, and it wasn't easy to adapt to the requirements of a new system. In short, conventional control panels proved to be very inflexible.

With invention of programmable (logic) controllers (PLC)[1], much has changed in how a process control system is designed. Many advantages appeared. Advantages of control panel that is based on a PLC controller can be presented in few basic points:

(1) Compared to a conventional process control system, number of wires needed for connections is reduced by 80%.

(2) Consumption is greatly reduced because a PLC consumes less than a bunch of relays.

(3) Diagnostic functions of a PLC controller allow for fast and easy error detection.

(4) Change in operating sequence or application of a PLC controller to a different operating process can easily be accomplished by replacing a program through a console or using a PC software (not requiring changes in wiring, unless addition of some input or output device is required).

(5) Needs fewer spare parts.

(6) It is much cheaper compared to a conventional system, especially in cases where a large number of I/O instruments are needed and when operational functions are complex.

(7) Reliability of a PLC is greater than that of an electro-mechanical relay or a timer.

Many PLC configurations are available; the configuration of the PLC refers to the packaging of the components. Typical configurations are listed below from largest to smallest (Fig 12.1).

 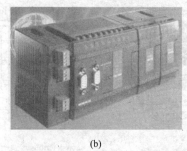 

(a) (b) (c)

Fig 12.1   PLC
(a) rack; (b) mini; (c) shoebox

Rack—A rack is often large and can hold multiple cards. When necessary, multiple racks can be connected together. These tend to be the highest cost, but also the most flexible and easy to maintain.

Mini—These are similar in function to PLC racks, but about half the size.

Shoebox—A compact, all-in-one unit (about the size of a shoebox) that has limited expansion capabilities. Lower cost, and compactness make these ideal for small applications.

Micro—These units can be as small as a deck of cards. They tend to have fixed

12.1.2   Basic components of PLC

PLC is actually an industrial microcontroller system (in more recent times we meet processors instead of microcontrollers) where you have hardware and software specifically adapted to industrial environment. Block schema with typical components which PLC consists of is found in Fig 12.2. Special attention needs to be given to input and output, because in these blocks you find protection needed in isolating a CPU blocks from damaging influences that industrial environment can bring to a CPU via input lines.

Central Processing Unit (CPU) is the brain of a PLC controller. CPU itself is usually one of the microcontrollers. A foretime these were 8-bit microcontrollers such as 8051, and now these are 16- and 32-bit microcontrollers. Unspoken rule is that you'll find mostly Hitachi[2] microcontrollers in PLC controllers by Japanese makers, Siemens[3] in European controllers, and Motorola[4] microcontrollers in American ones. CPU also takes care of communication, interconnectedness among other parts of PLC controller, program execution, memory operation, overseeing input and setting up of an output. PLC controllers have complex routines for memory checkup in order to ensure that PLC memory was not damaged (memory checkup is done for safety reasons). Generally speaking, CPU unit makes a great number of check-ups of the PLC controller itself so eventual errors would be discovered early. You can simply look at any PLC controller and see that there are several indicators in the form of light diodes for error signalization.

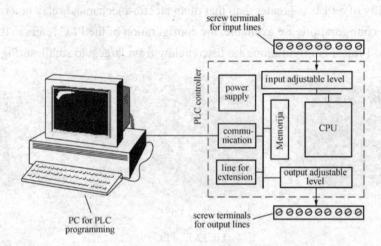

Fig 12.2　PLC's components

System memory (today mostly implemented in FLASH technology) is used by a PLC for an process control system. Aside from this operating system it also contains a user program translated from a ladder diagram to a binary form. FLASH memory contents can be changed only in case where user program is being changed. PLC controllers were used earlier have had EPROM memory instead of FLASH memory which had to be erased with UV lamp and programmed on programmers. With the use of FLASH technology this process was greatly shortened. Reprogramming a program memory is done through a serial cable in a program for application development.

User memory is divided into blocks having special functions. Some parts of a memory are used for storing input and output status. The real status of an input is stored either as "1" or as "0" in a specific memory bit. Each input or output has one corresponding bit in memory. Other parts of memory are used to store variable contents for variables used in user program. For example, timer value, or counter value would be stored in this part of the memory.

12.1.3　PLC Programming

The first PLCs were programmed with a technique that was based on relay logic wiring schematics. This eliminated the need to teach the electricians, technicians and engineers how to program a computer-but, this method has stuck and it is the most common technique for programming PLCs today. An example of ladder logic can be seen in Fig 12.3. To interpret this diagram we can imagine that the power is on the vertical line on the left hand side, we call this the hot rail. On the right hand side is the neutral rail. In the figure there are two rungs, and on each rung there are inputs or combinations of inputs (two vertical lines) and outputs (circles). If the inputs are closed in the right combination the power can flow from the hot rail, through the inputs, to power the outputs, and finally to the neutral rail. An input can come from a sensor, or switch. An output will be some device outside the PLC that is switched on or off, such as lights or motors. In the top rung the contacts are normally open and normally closed, which means if input A is on and input B is off, then power will flow through the output and activate it. Any other combination of input values will result in the output X being off.

The second rung of Fig 12.3 is more complex, there are actually multiple combinations of inputs that will result in the output Y turning on. On the left most part of the rung, power could flow through the top if C is off and D is on. Power could also (and simultaneously) flow through the bottom if both E and F are true. This would get power half way across the rung, and then if G or H is true the power will be delivered to output Y.

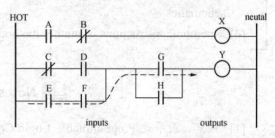

Fig 12.3   An example of ladder logic diagram

There are other methods for programming PLCs. One of the earliest techniques involved mnemonic instructions. These instructions can be derived directly from the ladder logic diagrams and entered into the PLC through a simple programming terminal. An example of mnemonics is shown in Fig 12.4. In this example the instructions are read one line at a time from top to bottom. The first line 0 has the instruction LDN (input load and not) for input 00001. This will examine the input to the PLC and if it is off it will remember a 1 (or true), if it is on it will remember a 0 (or false). The next line uses an LD (input load) statement to look at the input 00002. If the input is off it remembers a 0, if the input is on it remembers a 1. The AND statement recalls the last two numbers remembered and if they are both true the result is a 1, otherwise the result is a 0. The process is repeated for lines 00003 and 00004, the AND in line 5 combines the results from the last LD instructions. The OR instruction takes the two numbers now remaining and if either one is a 1 the result is a 1, otherwise the result is a 0. The last instruction is the ST (store output) that will look at the last value stored and if it is 1, the output will be turned on, if it is 0 the output will be turned off.

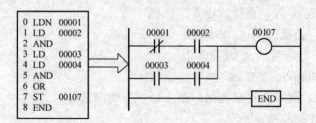

Fig 12.4   mnemonic & equivalent ladder logic program

The ladder logic program in Fig12.4, is equivalent to the mnemonic program. Even if you have programmed a PLC with ladder logic, it will be converted to mnemonic form before being used by the PLC.

Review

(1) PLC is actually an industrial microcontroller system where you have hardware and software specifically adapted to industrial environment.
(2) With invention of PLC, much has changed in how a process control system is designed.
(3) The first PLCs were programmed with a technique that was based on relay logic wiring schematics.
(4) Mnemonic instructions is another method for programming PLCs.

Notes to the text

[1] PLC 最早是 Programmable Logic Controller 的缩写，随着科技发展，PLC 的处理能力越来越强，内部采用 16 位，甚至 32 位微处理器，功能已不仅限于逻辑功能处理了，因而改称 Programmable Controller，但其缩写 PC 易与个人计算机缩写混淆，所以一般仍称 PLC。
[2] Hitachi　日本的日立公司
[3] Siemens　德国的西门子公司
[4] Motorola　美国的摩托罗拉公司

Technical Words

adapt [əˈdæpt] v. 改编，适应
console [kənˈsəul] n. [计] 控制台　vt. 安慰，慰问
diagnostic [daiəgˈnɔstik] adj. 诊断的
eliminate [iˈlimineit] vt. 排除，消除　v. 除去
inflexible [inˈfleksib(ə)l] adj. 不屈的，不屈挠的，顽固的
instrument [ˈistrum(ə)nt] n. 工具，手段，器械，器具，手段
oversee [əuvəˈsi:] v. 俯瞰，监视，检查，视察
panel [ˈpæn(ə)l] n. 面板，嵌板，仪表板　vt. 嵌镶板
rack [ræk] n. 架，行李架　vt. 放在架上
reliability [ri,laɪəˈbiləti] n. 可靠性
rung [rʌŋ] n. 地位，横档，这里指梯形图上的一条横向通路
vertical [ˈvɜ:tik(ə)l] adv. 垂直地

Technical Phrases

programmable (logic) controllers　　可编程控制器

microcontroller system	单片机系统
industrial environment	工业环境
control panel	控制面板
ladder diagram	梯形图
UV lamp	紫外灯

12.2 Reading materials

12.2.1 Sensors connected with PLC

Sensors allow a PLC to detect the state of a process. Logical sensors can only detect a state that is either true or false. Examples of physical phenomena that are typically detected are listed below.

(1) inductive proximity—is a metal object nearby?

(2) capacitive proximity—is a dielectric object nearby?

(3) optical presence—is an object breaking a light beam or reflecting light?

(4) mechanical contact—is an object touching a switch?

When a sensor detects a logical change it must signal that change to the PLC. This is typically done by switching a voltage or current on or off (Fig 12.5). Typical out-puts from sensors (and inputs to PLCs) are listed below:

(1) Plain Switches—Switches voltage on or off.

(2) Sinking/Sourcing sensor—Switches current on or off.

(3) Solid State Relays—These switch AC outputs.

(4) TTL (Transistor Transistor Logic)—Uses 0V and 5V to indicate logic levels.

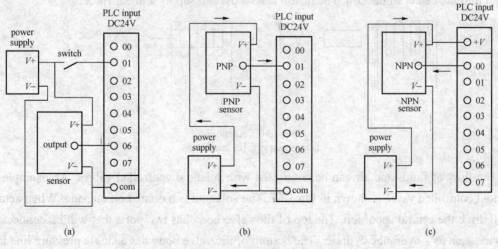

Fig 12.5   sensor wiring
(a) voltage signals; (b) sourcing sensors; (c) sinking sensors

Sinking sensors allow current to flow into the sensor to the voltage common, while sourcing sensors allow current to flow out of the sensor from a positive source. For both of these methods the emphasis is on current flow, not voltage. By using current flow, instead of voltage, many of the electrical noise problems are reduced.

When a PLC input card does not have a common but it has a V+ instead, it can be used for NPN sensors. In this case the current will flow out of the card (sourcing) and we must switch it to ground(Fig 12.5).

12.2.2 Actuators

Actuators drive devices in mechanical systems. Most of them are by converting electrical energy into some form of mechanical motion. There are many types of actuators including those on the brief list below:

(1) Solenoids can be used to convert an electric current to a limited linear motion.

(2) Solenoid valves can be used to redirect fluid and gas flows.

(3) Heaters—They are often controlled with a relay and turned on and off to maintain a temperature within a range.

(4) Lights—Lights are used on almost all machines to indicate the machine state and provide feedback to the operator. most lights are low current and are connected directly to the PLC.

(5) Hydraulics and pneumatics use cylinders to convert fluid and gas flows to limited linear motions. Pneumatics provides smaller forces at higher speeds, but is not stiff. Hydraulics provides large forces and is rigid, but at lower speeds.

Here take solenoids as an example. Solenoids are the most common actuator components. The basic principle of operation is there is a moving ferrous core (piston) that will move inside wire coil as shown in Fig 12.6. Normally the piston is held outside the coil by a spring. When a voltage is applied to the coil and current flows, the coil builds up a magnetic field that attracts the piston and pulls it into the center of the coil. The piston can be used to supply a linear force.

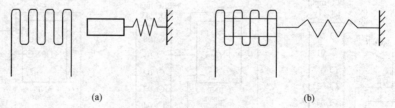

Fig 12.6 solenoids
(a) current off; (b) current on

The flow of fluids and air can be controlled with solenoid controlled valves. An example of a solenoid controlled valve is shown in Fig 12.7. The solenoid is mounted on the side. When actuated it will drive the central spool left. The top of the valve body has two ports that will be connected to a device such as a hydraulic cylinder. The bottom of the valve body has a single pressure line in the center with two exhausts to the side. In Fig 12.7(a) drawing the power flows in through the center to the right hand cylinder port. The left hand cylinder port is allowed to exit through an exhaust port. In Fig 12.7(b) drawing the solenoid is in a new position and the pressure is now applied to the left hand port on the top, and the right hand port can exhaust. Valves are also available that allow the valves to be blocked when unused.

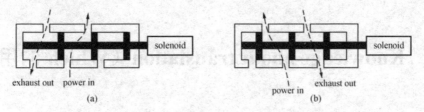

Fig 12.7　solenoid controlled valve
(a) exhaust out form left port; (b) exhaust out from right port

12.2.3　PLC compared with other control systems

PLCs are well-adapted to a range of automation tasks. These are typically industrial processes in manufacturing where the cost of developing and maintaining the automation system is high relative to the total cost of the automation, and where changes to the system would be expected during its operational life. PLCs contain input and output devices compatible with industrial devices and controls; little electrical design is required, and the design problem centers on expressing the desired sequence of operations in ladder logic (or function chart) notation.

PLC applications are typically highly customized systems so the cost of a packaged PLC is low compared to the cost of a specific custom-built controller design. On the other hand, in the case of mass-produced goods, customized control systems are economic due to the lower cost of the components, which can be optimally chosen instead of a "generic" solution. For high volume or very simple fixed automation tasks, different techniques are used. For example, a consumer dishwasher would be controlled by an electromechanical cam timer costing only a few dollars in production quantities.

A microcontroller-based design would be appropriate where hundreds or thousands of units will be produced and so the development cost (design of power supplies and input/output hardware) can be spread over many sales, and where the end-user would not need to alter the control. Automotive applications are an example; millions of units are built each year, and very few end-users alter the programming of these controllers. However, some specialty vehicles such as transit busses economically use PLCs instead of custom-designed controls, because the volumes are low and the development cost would be uneconomic.

Very complex process control, such as used in the chemical industry, may require algorithms and performance beyond the capability of even high-performance PLCs. Very high-speed or precision controls may also require customized solutions; for example, aircraft flight controls.

PLCs may include logic for single-variable feedback analog control loop, a "proportional, integral, derivative" or "PID controller". A PID loop could be used to control the temperature of a manufacturing process, for example. Historically PLCs were usually configured with only a few analog control loops; where processes required hundreds or thousands of loops, a distributed control system (DCS) would instead be used. However, as PLCs have become more powerful, the boundary between DCS and PLC applications has become less clear-cut.

12.3 Knowledge about translation（Which 的用法）

1. 用作关系代词

Which 用作关系代词时，引导出定语从句。这时它不仅起到把主句与从句连接起来的作用，而且还用来代替从句中某一名词。Which 在从句中常用作主语、宾语或与介词连用起状语作用。

以前的语法书中 Which 引导定语从句有限定性定语从句和非限性定语从句之分，但目前这种区分日益淡化，所以本书中也不加区分。但现在一般用法，which 引导的从句都用逗号与主句分开，便于阅读，翻译时可根据需要在从句前加上"因为""如果""由于""虽然"等相应的词。

A conductor is a substance, which is able to carry electrons easily.

导体是能很容易传输电子的物质。

Electrical engineering has many sub-disciplines, the most popular of which are listed below.

电气工程有很多子学科，以下列出其中最常开设的一些子学科。which 代替 sub-disciplines。

It develops an ac resistance, known as capacitive reactance, which is affected by the capacitance and ac frequency.

电容对交流电有阻抗作用，称为容抗，容抗与电容值和所加交流电的频率有关。Which 引导的非限制性定语从句，进一步说明 reactance。

The process by which we determine a variable (either voltage or current) of a circuit is called analysis.

求出电路的变量（电压或电流）的过程称为（电路）分析。by which 连用起状语作用。

有时 Which 引导的特种定语从句通常用来代表主句的内容或部分内容，which 在定语从句中常常作主语，也可作定语或介词的宾语。作主语时谓语动词要用第三人称单数。

You introduce some negative feedback into the signal path, which reduces distortion.

把负反馈引回到信号通道，这样减少（输出）失真。

When the motor is turned on, a current flows in the wire, which creates a sizable magnetic field around the wire coil.

当电动机转动时，导线中就有电流流动，这就产生了一个环绕着绕组线圈的相当大的磁场。which 指 a current flows in the wire。

2. "介词+which" 引导的定语从句

介词与 Which 连用，在所引导定语从句中构成介词短语，主要用作状语。介词属于从句，分析时应当从介词开始断开。

In analysis of electrical circuits, there are several distinct approaches that we can take. In the one we write a set of simultaneous equations in which the variables are voltage; this is known as nodal analysis.

分析电路有几种方法，其中一种是列出同一时刻的方程组，其变量是电压，这种方法称为节点（电位）分析法。这里 which 指前面的 a set of simultaneous equations。

System-on-a-Chip (SoC or SOC) is an integrated circuit in which all the components needed for a computer or other system is included on a single chip.

单芯片系统（SoC 或 SOC）是把一个计算机或其他系统所需要的全部器件全集成在单一芯片上的集成电路。in which 处断开，which 指 an integrated circuit。

The enclosure also protects the electrical and operating parts of the motor from harmful effects of the environment in which the motor operates.

当电动机在比较恶劣的环境中工作时，外壳可以保护电动机的电气和运动部分。这里 which 指 the environment。

There are different forms of energy around us, any form of which can be changed into another form. 我们周围存在着不同形式的能量，其中任何形式的能量都能转变为另一种形式。这里的 which 指代 energy。

The furnace contains a complex of tubes and drums, called a boiler, through which water is pumped and the temperature of the water rises in the process until the water evaporates into steam.

锅炉中有复杂的管道和鼓状的设备，称为沸腾器，水泵把水压入沸腾器，在沸腾器中水温度上升并蒸发成为蒸汽。句子比较长，可以把它译成几句话。

3. Which 用作连接代词

Which 用作连接代词时，保留疑问含义，可引导主语从句，表语从句和宾语从句。通常译成"哪个""哪些"。

Which process will occur is determined entirely by the surroundings. 将出现哪种过程，完全取决于周围的环境。

4. Which 用在强调句型中

It is (was) + 强调成分+which 这一强调句型中，只能强调句中的主语和宾语，但不能用来强调句中的状语。这种用法现在不多。

12.4　Exercises

1. Put the Phrases into English (将下列词组译成英语)
(1) 工业革命
(2) 可编程控制器
(3) 控制面板
(4) PLC 的可靠性
(5) 硬件和软件

(6) 闪存技术

(7) 应用开发

(8) 输出状态

(9) 梯形图

(10) 多重组合

2. Put the Phrases into Chinese (将下列词组译成中文)

(1) covered an entire wall

(2) process control systems

(3) Needs fewer spare parts

(4) a large number of I/O instruments

(5) the packaging of the components

(6) interconnectedness among other parts of PLC controller

(7) error signalization

(8) through a serial cable

(9) the real status of an input

(10) one of the earliest techniques

3. Sentence Translation (将下列句子译成中文)

(1) To discover an error in the system much time was needed especially with more complex process control systems.

(2) It is much cheaper compared to a conventional system, especially in cases where a large number of I/O instruments are needed and when operational functions are complex.

(3) CPU also takes care of communication, interconnectedness among other parts of PLC controller, program execution, memory operation, overseeing input and setting up of an output.

(4) For example, timer value, or counter value would be stored in this part of the memory.

(5) These instructions can be derived directly from the ladder logic diagrams and entered into the PLC through a simple programming terminal.

4. Translation (翻译)

Unlike general-purpose computers, the PLC is designed for multiple inputs and output arrangements, extended temperature ranges, immunity to electrical noise, and resistance to vibration and impact. Programs to control machine operation are typically stored in battery-backed or flash memory. A PLC is an example of a real time system since output results must be produced in response to input conditions within a bounded time.

12.5　课文参考译文

可编程逻辑控制器，简称 PLC 或可编程控制器是一个数字计算机设备，主要用于工业机电设备工作过程的自动化控制，如机器的控制、工厂装配生产流水线的控制、游乐设施的控制或灯光照明设备的控制。PLC 设计成有多组数字信号和模拟信号的输入输出端子，工作温度范围比较宽，抗电噪音能力强，抗震、防碰撞性能好。一个 PLC 是一个典型的"硬件"实时系统，因为对输入信号必须在很短时间里产生输出的响应，否则会产生误操作。

12.5.1 PLC 简介

作为工业革命的产物，尤其在 20 世纪六七十年代，机器的自动化运行主要用继电器，这些继电器用导线相互连接，安放在控制柜中（对机器进行控制的）。有时一个控制柜有一面墙那么大。在这样的（控制）系统尤其是比较复杂的过程控制系统中要找出一个错误需要很多时间。一个最主要的问题是继电器的触点的寿命有限，有时需要更换继电器。如果要更换继电器时，机器和生产过程都必须停下来，有时甚至没有足够的地方来更换继电器。另外控制柜只能用于一个特殊的生产过程，控制柜不能很容易适应一个新的系统的控制要求（控制要求变化时，继电器的种类、数量和连接方式等都要改变）。简单地说传统继电控制柜的适应性不强。

随着可编程（逻辑）控制器（PLC）的发明，控制系统的设计过程发生了很多变化。出现了许多新的技术。用一个 PLC 来取代控制柜主要的好处如下：

（1）与传统的控制柜系统比较，所需的连接线减小了 80%。

（2）因为 PLC 的损耗小于一大堆继电器，所以损耗减少了很多。

（3）PLC 的诊断功能使得比较容易快速找到系统的错误。

（4）可以通过编程器或用 PC 软件换一段程序就可以很方便地改变（系统的）操作顺序或 PLC 在不同操作过程中的应用（除非要增加一些的输入输出设备，否则不需要改变连接线）。

（5）只需要较少的零部件。

（6）与传统的（控制）系统相比，采用 PLC 比较便宜，尤其是需要大量输入输出仪器（设备）和操作功能比较复杂的场合。

（7）PLC 可靠性比电磁继电器或时间继电器高。

PLC 有多种结构，PLC 的结构是指 PLC 部件的封装形式，下面列出从最大到最小的典型结构（图 12.1）。

（架子）组合式——架子通常比较大，可以容纳多个 PLC 模块（卡），需要时，还可以把多个架子连接在一起，这样成本最高，但适应性最强，容易维护。

小型——功能上与架子组合很像（也可以拼接），但一般比它小一半。

单体式（鞋盒）——所有功能合在一个单元里（约和鞋盒一样大）的紧凑型 PLC，有限的扩展功能，成本低，因其体积小，在一些小的应用场合使用比较理想。

微型——体积小到只有一副扑克牌那么大，一般都是固定的（不能扩展）。

12.5.2 PLC 的基本部件

PLC 实际上是一个工业微控制器系统（近年来已经用微处理器取代了微控制器）其中有特别设计可以适应工业环境的硬件系统和软件系统。PLC 的典型构成部件如框图 12.2 所示。特别值得一提的是输入和输出（模块），因为在这些模块里有隔离保护措施，可以隔离工业环境通过输入带给 CPU 的不良影响。

中央处理单元（CPU）是 PLC 的大脑，CPU 本身是用一种微控制器，早期是 8 位的处理器（如 8051），现在有 16 位和 32 位的微处理器。默认的是在日本制造的 PLC 中用得较多的是日立微控制器，在欧洲制造的 PLC 中用得较多的是西门子的微控制器，美国制造的 PLC 中用得较多的是摩托罗拉的微控制器。CPU 还要处理通信，和 PLC 的其他部件的相互连接，执行程序，对存储器进行操作，监视输入端口和设置输出端口（输出信号）。PLC 控制器还有较复杂的存储器常规检查，以确保 PLC 的存储器没有损坏（存储器检查是出于安全性的考虑）。一般 CPU 承担了大量的对 PLC 本身的检查工作，所以可以较早地发现错误。可以发现任何一个 PLC 都有一些发光二极管作为指示灯用作出错信号（报警）。

PLC 把控制处理过程存储在系统的存储器（目前大部分是用闪存技术实现的）。除了操作系统外，存储器中还储存用户所编写的已从梯形图转换成二进制代码的程序。只有当用户改变程序时闪存的内容才会改变。早期的 PLC 是用 EPROM 来作为存储器的，（重新）编程时要先用紫外光灯擦除（原来存储的程序），才可以写入新的程序。采用了闪存技术后重新编程这个过程就简化了，只要在开发应用软件中重新编程然后用串行电缆输入就可以了。

用户的存储器分成几块各有指定的功能，一部分存储单元是用于存储输入和输出的状态的，一个输入的实际状态是在一个指定的二进制位单元中存储一个 1 或一个 0。每个输入或输出在存储器中都有相应的一个指定的二进制位。存储器的其他部分是用于存储用户程序中所用到的变量的值，例如，时间值或计数值都可以存储在这部分单元中。

12.5.3　PLC 编程

最初 PLC 编程是采用与继电器逻辑原理图相同的技术（梯形图），这样电工、技术员和工程师都不需要学习计算机编程，但是这种方法被普遍接受了，并成为今天 PLC 编程的最通用的技术。图 12.3 给出一个梯形图，为了说明这个图，可以想象左边的竖线为电源线，称为火线，右边的是零（中）线。图中有两条通路，每一条通路上都有输入或输入的组合（两条水平线）和输出（圆形符号），如果通过适当组合使输入是闭合的，电流就可以从火线通过输入，流过输出最终到零线。输入可以来自传感器或开关。输出可以是 PLC 外接的一些器件，它控制如灯或电动机的通电或断开。在第一条通路中连接一个动合触点和一个动断触点，如果输入 A 是合上（则 A 动作，其常开触点成为闭合），B 是断开（则 B 不动作，其常闭触点仍为闭合），则电流会流过输出并使输出有效。而任何其他输入的组合（另三种分别为：A 合上，B 合上；A 断开，B 合上；A 断开，B 断开）都会使（第一条通路不通）输出 X 是断开。

图 12.3 的第二条通路比较复杂，有多个可能的输入组合会使输出 Y 接通。在通路左部，上面一条如果 C 断开、D 接通电流可以流过，下面一条，如果 E 和 F 都接通电流也可以流过，这时电流流过半条通路，如果还有 G 或 H 接通则电流就流入输出端 Y 了。

PLC 编程还有其他方法，其中最早的技术之一是符号指令。这些指令可以直接从梯形（逻辑）图中导出并通过一个编程引脚输入到 PLC（的存储器）中。图 12.4 是符号指令的一个例子，在这个例子中，可以从顶部到底部，每次读一条通路得到符号指令。第 0 行是对输入端 00001 的指令 LDN（输入加载并取反）。这条指令会检查输入端 00001，如果是断开就读入输入 0 并取反，输给 PLC 一个 1（或真），如果输入端接通就输给 PLC 一个 0（或假）。下一条用 LD（输入加载）语句来检测输入端 00002，如果输入端是断开就记为 0，如果输入为通电，则记为 1。AND 语句是把上二条语句的值相"与"，如果都为真则结果为 1，否则结果为 0。对 00003 和 00004 的输入处理过程相同，第 5 行 AND 指令把后两个 LD 指令的输出值相"与"，得到一个输出值，OR 指令对保留的两个值进行"或"运算，如果任何一个为 1，输出为 1。否则结果为 0。最后的指令是 ST（存储输出），会把最终的输出值存储起来，如果是 1 则输出端为通电，如果是 0 则输出端为断开。

图 12.4 的梯形图与代码程序等效，即使你是用梯形图编程的，也要把它转换成代码形式输给 PLC，PLC 才能执行。

12.6 阅读材料参考译文

12.6.1 与 PLC 连接的传感器

PLC 用传感器去检测一个（工作）过程的状态，逻辑量（位）传感器只能检测一个状态是真（条件成立）还是假（条件不成立）。典型可检测到的物理现象的例子如下：

（1）电感——有金属物体靠近？
（2）电容——有绝缘物体靠近？
（3）存在光线——有物体切断光线或反射光线？
（4）机械接触——有物体撞到了开关？

当一个传感器检测到一个逻辑量的变化，它会把这个变化送给 PLC，通常是用一个电压的高低电平或者电流的流动或断开（图 12.5）来表示。典型的传感器的输出（也是 PLC 的输入）如下：

（1）普通开关——切换电压。
（2）接收/输出传感器——接通或断开电流。
（3）固态继电器——切换交流输出或不输出。
（4）TTL（晶体管传递逻辑）——用 0V 或 5V 表示逻辑电平。

接收（灌电流）传感器是让电流流入传感器再到电压公共端，而输出（拉电流）传感器是电流从正电源经传感器流出。这两种传感器关心的都是电流的流动，而不是电压。用这种电流信号取代电压（作为传感器信号）可以克服许多电噪声问题。

如果一个 PLC 输入部分没有公共端但有正电压端，可以用 NPN 传感器（图 12.5），这时电流会从 PLC 的电源流出，要把它接地。

12.6.2 执行器

在机械系统中执行器驱动设备，大部分执行器把电能转换成机械运动的形式，有很多种执行器，以下简单列出几种：

（1）螺线管可用于把电流转换成有限的线性运动（位移或角位移）。
（2）电磁阀可用于控制液体和气体的流动。
（3）加热器——通常用继电器控制加热器，使其在一定范围内维持温度。
（4）灯——几乎所有的机器都用灯表示其工作状态（反馈给操作者）。很多灯是用低电流的（耗电量小），直接连接在 PLC 的输出端。

（5）液压缸和气压缸用于把液体和气体流动转换成有限的线性运动。气压缸输出的力比较小，速度比较快，比较灵活；液压缸输出力较大，比较刚性，但速度比较慢。

这里以螺线管为例。螺线管是最常用的执行元件。基本工作原理是在线圈里有一个可移动的铁芯核（活动棒或活塞），它可以在线圈内移动，如图 12.6 所示。不通电时弹簧把活动棒拉出，在线圈的外面。当线圈加上电压时，电流流过，线圈中产生磁场，吸引活动棒，把它拉入线圈内，这个活动棒用于提供一个线性力。

可以用螺线管控制的阀（电磁阀）控制液体和气体的流动，图 12.7 给出一个电磁阀的例子。螺线管装在边上，当激励它（给它通电）时它驱动中心的（铁芯）轴移动，阀体的顶部有两个口，可以与如液压缸之类的设备相连接。阀的底部中心有一个压力（泵），两边有两个排放口。当铁芯移动到图 12.7（a）所示位置时，压力泵把能量（液体或气体）通过中心压入到右边的缸口，左边的液压缸口则通过一个排放口排放。当螺线管驱动铁芯到一个新的位置如图 12.7（b）所示时，现在压力泵把能量（液体或气体）通过中心压入到顶部的左边缸口，右边缸口则通过另一个排放口排放。在不用时电磁阀也可以断开液体（气体）流。

12.6.3　PLC 与其他控制系统的比较

PLC 可以很好地满足各种自动化任务的需要，制造业生产过程的典型特点是相对于自动化系统的总成本来说，开发和维护自动化系统的成本是相当高的，而且在自动化系统的运行周期中很可能要对系统作一些调整。PLC 含有与工业器件和控制相匹配的输入输出器件，因此在设计自动化控制系统时只需要作少量的电气设备方面的设计，主要的设计集中在用梯形图（或功能图）表示所想要的操作流程上。

PLC 应用通常是高度定制化的系统（适用于自动化过程比较复杂且有时需要调整自动化流程），所以相对于专用定制控制系统来说用现成的 PLC 成本比较低。反过来说，如果是大批量生产的商品（所需的控制系统），用户定制的控制系统由于采用较低成本的零件会比较经济，因此用户可以（根据需要）优化选取自动化控制系统。对大量或者十分简单的自动化任务，可以用不同的技术。例如，一个洗碗机可以用一个机电凸轮定时器来控制，批量生产时一个机电凸轮定时器成本只有几元钱。

基于微控制器的设计比较适于要生产几百或几千个（控制）单元，因此开发成本（电源的设计，输入/输出口硬件）会分摊到许多产品上且用户不需要改变控制过程。汽车产品就是一个例子，每年要生产上万个控制单元且几乎没有终端用户要改变这些控制程序。但是，一些特别的车辆如公交车用 PLC 代替批量设计的控制器比较经济，因为定制控制器数量少，开发成本就比较高。

非常复杂的控制过程，如在化学工业中所用的控制过程，所需要的算法和控制性能可能连高性能的 PLC 也无法实现，需要定制控制系统。非常高速或高精度的控制（如飞机飞行控制），也需要定制控制方案。

PLC 可能包含单变量反馈模拟闭环控制，"比例积分微分"控制或称 PID 控制器的功能，例如，一个 PID 闭环可用于控制制造过程中的温度。以前 PLC 一般只可以设置几个模拟控制闭路环，如果过程中需要上百个或上千个这样的模拟控制闭路环，就要用集散控制系统（DCS）来代替 PLC。但是随着 PLC 的功能越来越强大，集散控制系统 DCS 和 PLC 的应用界线就变得不明显了。

Unit 13　Automation

13.1　Text

Automation or industrial automation is the use of control systems such as computers or PLCs to control industrial machinery and processes, reducing the need for human intervention. Automation plays an increasingly important role in the global economy. Engineers strive to combine automated devices with mathematical and organizational tools to create complex systems for a rapidly expanding range of applications and human activities. Here have some simple examples of automation.

13.1.1　Automation of parking garage

We are dealing with a simple system that can control 100 cars at the maximum (Fig 13.1). Each time a car enters, PLC automatically adds it to a total sum of other cars found in the garage. Each car that comes out will automatically be taken off. When 100 cars park, a signal will turn on signaling that a garage is full and notifying other drivers not to enter because there is no space available.

Signal from a sensor at the garage entrance sets bit IR200.00. This bit is a condition for execution of the following two instructions in a program. First instruction resets carry bit CY (it is always done before some other calculation that would influence it), and the other instruction adds one to a number of cars in word HR00, and a sum total is again stored in

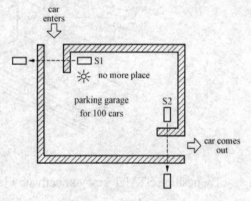

Fig 13.1　Car garage

HR00. HR00 memory space is selected for storing a total number of cars because this keeps the status even after supply stops. Symbol in addition and subtraction instructions defines decimal constant that is being added or subtracted from a number of cars already in the garage. Condition for executing comparison instruction CPM is always executed because bit SR253.13 is always set; this practically means that comparison will be done in each cycle regardless whether car has entered or left the garage. Signal lamp for "garage full" is connected to an output IR010.00. Working of the lamp is controlled by EQ (equal) flag at address SR255.06 and GR (greater than) flag at address.

Both bits are in OR connection with an output where the signal lamp is. This way lamp will emit light when a number of cars is greater than or equal to 100. Number of cars in a real setting can really be greater than 100 because some untrusting driver may decide to check whether there is any space left, and so a current number of cars can increase from a 100 to 101. When he leaves the garage, a number of cars goes down to 100 which is how many parking spots there are in fact.

13.1.2 Automation of product packaging

Product packaging is one of the most frequent cases for automation in industry. It can be encountered with small machines (ex. packaging grain like food products) and large systems such as machines for packaging medications. Example we are showing here solves the classic packaging problem with few elements of automation (Fig 13.2). Small number of needed inputs and outputs provides for the use of CPM1A PLC controller.

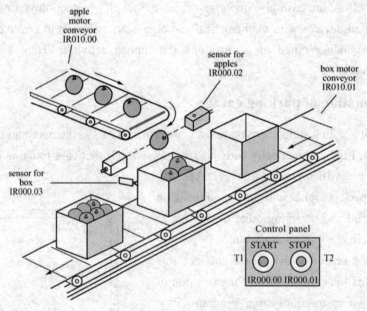

Fig 13.2 a packaging line

By pushing START key you activate Flag1 which represents an assisting flag (Segment 1) that comes up as a condition in further program (resetting depends only on a STOP key). When started, motor of a conveyor for boxes is activated. The conveyor takes a box up to the limit switch, and a motor stops then (Segment 4). Condition for starting a conveyor with apples is actually a limit switch for a box. When a box is detected, a conveyor with apples starts moving (Segment 2). Presence of the box allows counter to count 10 apples through a sensor used for apples and to generate counter CNT010 flag which is a condition for new activation of a conveyor with boxes (Segment 3). When the conveyor with boxes has been activated, limit switch resets counter which is again ready to count 10 apples. Operations repeat until STOP key is pressed when condition for setting Flag1 is lost. Picture in Fig 13.3 gives a time diagram for a packaging line signal.

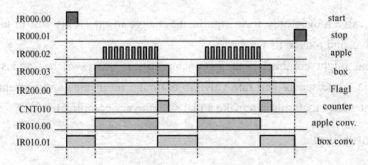

Fig 13.3   a time diagram for a packaging line signal

13.1.3   Water tank level control

A facility needs to store water in a tank. The water is drawn from the tank by another system, as needed, and system must manage the water level in the tank. There are two kind of signals to control, digital and analog.

Using only digital signals, the PLC has two digital inputs from switches (tank empty and tank full). The PLC uses a digital output to open and close the water supply valve into the tank. The ladder diagram of control process is shown in (Fig 13.4).

When the water level drops enough so that the tank empty switch is on (up), the PLC will open the valve to let more water in. Once the water level raises enough so that the tank full switch is off (down), the PLC will shut the valve to stop the water from overflowing.

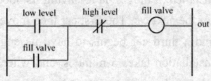

Fig 13.4   ladder diagram of control process

An analog system (Fig 13.5) might use a water pressure sensor or a load cell, and an adjustable (throttling)dripping out of the tank, the valve adjusts to slowly drip water back into the tank.

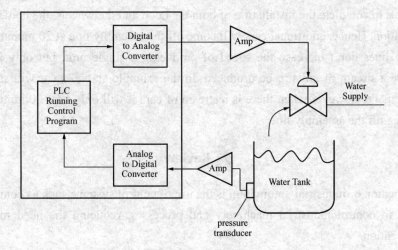

Fig 13.5   Water Tank Level Controller

In this system, to avoid 'flutter' adjustments that can wear out the valve, many PLCs incorporate "hysteresis" which essentially creates a "deadband" of activity. A technician adjusts this deadband so the valve moves only for a significant change in rate. This will in turn minimize the motion of the valve, and reduce its wear.

A real system might combine both approaches, using float switches and simple valves to prevent spills, and a rate sensor and rate valve to optimize refill rates and prevent water hammer. Backup and maintenance methods can make a real system very complicated.

13.1.4 Assembly line

An assembly line is a manufacturing process in which parts are added to a product in a sequential manner using optimally planned logistics to create a finished product much faster than with handcrafting-type methods. The best known form of the assembly line, the moving assembly line, was realized into practice by Ford Motor Company between 1908 and 1915. Mass production via assembly lines is widely considered to be the catalyst which initiated the modern consumer culture by making possible low unit costs for manufactured goods.

Consider the assembly of a car: assume that certain steps in the assembly line are to install the engine, install the hood, and install the wheels (in that order, with arbitrary interstitial steps). A car on the assembly line can have only one of the three steps done at once. After the car has its engine installed, it moves on to having its hood installed, leaving the engine installation facilities available for the next car. The first car then moves on to wheel installation, the second car to hood installation, and a third car begins to have its engine installed. If engine installation takes 20 minutes, hood installation takes 5 minutes, and wheel installation takes 10 minutes, then finishing all three cars when only one car can be operated at once would take 105 minutes.

On the other hand, using the assembly line, the total time to complete all three cars is 75 minutes. This is possible because there are 3 different installation stations, an engine station, a hood station and a wheels station. Because it takes 20 minutes to finish work on the engine while it takes only 15 minutes to complete the installation of both the hood and the wheels, the bottleneck is at the engine installation. Hence, additional cars will come off the assembly line at 20 minute increments.

Assembly lines don't increase the speed of producing a single unit, but only increases rate when there are a stream of units to be produced. In the example above, a car will still require 35 minutes to be made, however when there is a stream of cars it will only take 20 minutes to have a new car coming off the assembly line.

 Review

(1) Automation or industrial automation is the use of control systems such as computers or PLCs to control industrial machinery and processes, reducing the need for human intervention.

(2) An assembly line is a manufacturing process in which parts are added to a product in a sequential manner using optimally planned logistics to create a finished product much faster than with handcrafting-type methods.

Technical Words

approach [əˈprəʊtʃ] n. 接近，逼近，方法，步骤 vt. 接近，动手处理 vi. 靠近
arbitrary [ˈɑːbɪt(rə)ri] adj. 任意的，武断的，独裁的，专断的
automation [ɔːtəˈmeɪʃ(ə)n] n. 自动控制，自动操作
catalyst [ˈkæt(ə)lɪst] n. 催化剂
conveyor [kənˈveɪə] n. 传送，运输，传送带，输送机
drip [drɪp] n. 水滴 v. （使）滴下
facility [fəˈsɪləti] n. 容易，简易，灵巧，熟练，便利，敏捷，设备，工具
flag [flæɡ] n. 旗，标记 v. 做标记
flutter [ˈflʌtə] n. 摆动，鼓翼 vi. 鼓翼，飘动，波动
garage [ˈɡærɑː(d)ʒ] n. 汽车间，修车厂，车库 v. 放入车库
hood [hʊd] n. 引擎罩
hysteresis [ˌhɪstəˈriːsɪs] n. 滞后作用，[物]磁滞现象
influence [ˈɪnfluəns] n. 影响，感化，（电磁）感应 vt. 影响，改变
interstitial [ˌɪntəˈstɪʃ(ə)l] adj. 空隙的，裂缝的，形成空隙的 n. 插屏广告
intervention [ɪntəˈvenʃ(ə)n] n. 干涉
maintenance [ˈmeɪnt(ə)nəns] n. 维护，保持，生活费用，扶养
package [ˈpækɪdʒ] n. 包裹，包装
part [pɑːt] n. 部分，局部，零件 vt. 分开，分离，分配
refill [riːˈfɪl] v. 再装满，补充，再充填 n. 新补充物，替换物
spill [spɪl] n. 溢出，溅出 vt. 使溢出，使散落，洒 vi. 溢出，充满
tank [tæŋk] n. 桶，箱，罐，坦克 vt. 储于槽中
throttle [ˈθrɒt(ə)l] v. 扼杀 throttling 扼流

Technical Phrases

water hammer	水锤
assembly line	流水线，装配线
handcrafting-type method	手工方式
mass production	批量生产
rate valve	速率阀
consumer culture	消费文化
manufactured good	制造商品
wear out	磨损
deadband	死区

13.2 Reading materials

13.2.1 VFD system

A variable-frequency drive (VFD) is a system for controlling the rotational speed of an alternating current (AC) electric motor by controlling the frequency of the electrical power supplied to the motor. A variable-frequency drive is a specific type of adjustable-speed drive. Variable-frequency drives are also known as adjustable-frequency drives (AFD), variable-speed drives (VSD), AC drives, microdrives or inverter drives. Since the voltage is varied along with frequency, these are sometimes also called VVVF (variable voltage variable frequency) drives.

A variable frequency drive system generally consists of an AC motor, a VFD controller and an operator interface. (Fig 13.6).

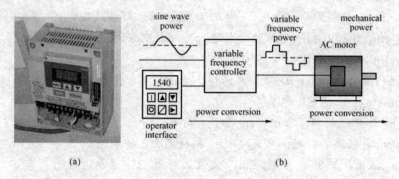

Fig 13.6 VFD & VFD system
(a) VFD controller; (b) VFD system

The motor used in a VFD system is usually a three-phase induction motor. Some types of single-phase motors can be used, but three-phase motors are usually preferred. Various types of synchronous motors offer advantages in some situations, but induction motors are suitable for most purposes and are generally the most economical choice. Motors that are designed for fixed-speed mains voltage operation are often used, but certain enhancements to the standard motor designs offer higher reliability and better VFD performance.

Variable frequency drive controllers are solid state electronic power conversion devices. The usual design first converts AC input power to DC intermediate power using a rectifier bridge. The DC intermediate power is then converted to quasi-sinusoidal AC power using an inverter switching circuit. The rectifier is usually a three-phase diode bridge, but controlled rectifier circuits are also used (Fig 13.7).

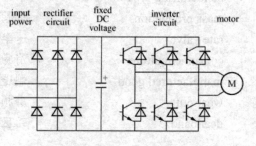

Fig 13.7 principle of VFD

The operator interface provides a means for an operator to start and stop the motor and adjust the operating speed. Additional operator control functions might include reversing and switching between manual speed adjustment and automatic control from an external process control signal. The operator interface often includes an alphanumeric display and/or indication lights and meters to provide information about the operation of the drive.

13.2.2 Robot & its control

A typical robot (Fig 13.8) will have several, though not necessarily all of the following properties:

(1) is not "natural", i.e. it is artificially created.

(2) can sense its environment, and manipulate or interact with things in it.

(3) has some ability to make choices based on the environment, often using automatic control or a preprogrammed sequence.

(4) moves with one or more axes of rotation or translation.

(5) appears to have intent or agency.

The mechanical structure of a robot must be controlled to perform tasks. The control of a robot involves three distinct phases-perception, processing and action. Sensors give information about the environment or the robot itself (e.g. the position of its joints or its end effector). This information is then processed to calculate the appropriate signals to the actuators (motors) which move the mechanical structure.

Fig 13.8   robot

The processing phase can range in complexity. It may be the first step to estimate parameters of interest (e.g. the position of the robot's gripper) from noisy sensor data. An immediate task (such as moving the gripper in a certain direction) is inferred from these estimates. Techniques from control theory convert the task into commands that drive the actuators.

At longer time scales or with more sophisticated tasks, the robot may need to build and reason with a "cognitive" model. Cognitive models try to represent the robot, the world, and how they interact. Pattern recognition and computer vision can be used to track objects. Mapping techniques can be used to build maps of the world. Finally, motion planning and other artificial intelligence techniques may be used to figure out how to act.

Control systems may also have varying levels of autonomy. Direct interaction is used for haptic or tele-operated devices, and the human has nearly complete control over the robot's motion. Operator-assist modes have the operator commanding medium-to-high-level tasks, with the robot automatically figuring out how to achieve them. An autonomous robot may go for extended periods of time without human interaction. Higher levels of autonomy do not necessarily require more complex cognitive capabilities. For example, robots in assembly plants are completely autonomous, but operate in a fixed pattern.

13.2.3   Process control system

Generally speaking, process control system is made up of a group of electronic devices and equipment that provide stability, accuracy and eliminate harmful transition statuses in production processes.

Operating system can have different form and implementation, from energy supply units to machines. As a result of fast progress in technology, many complex operational tasks have been solved by connecting programmable logic controllers and possibly a central computer. Beside connections with instruments like operating panels, motors, sensors, switches, valves, possibilities for communication among instruments are so great that they allow high level of exploitation and process coordination, as well as greater flexibility in realizing a process control system. Each component of a process control system plays an important role, regardless of its size. For example, without a sensor, PLC wouldn't know what exactly goes on in the process. In automated system, PLC controller is usually the central part of an process control system. With execution of a program stored in program memory, PLC continuously monitors status of the system through signals from input devices. Based on the logic implemented in the program, PLC determines which actions need to be executed with output instruments. To run more complex processes it is possible to connect more PLC controllers to a central computer. A real system could look like the one pictured in Fig 13.9.

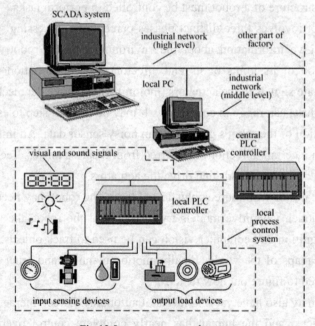

Fig 13.9　process control system

13.3　Knowledge about translation（倒装）

英语句子与中文一样，通常是主语在前，谓语其次，宾语在后，这种语序称为自然语序，否则称为倒装语序。使用倒装语序通常是为了强调句中的某个成分，或者是由于句子结构的要求，如英语的问句一般为倒装句英语一些习惯用法也有用倒装句的。

一般的倒装句如疑问句和 there be 句型，这里不再讨论。这里讨论科技英语中常用且在翻译时容易出错的倒装句型。

1. 句首有表示否定的词

在英语中，当一些含有否定意义的词放在句首时，一般要用动词在前，主语在后的倒装结构，这些词有：not, no, never, hardly（几乎不），seldom（很少），scarcely（几乎不），not until, not only, neither, nor , in no way, no longer（不再），nowhere 等。

Not only **can** electricity be made to produce magnetism, but magnetism can be made to produce electricity.

不但电可以产生磁，磁也可以产生电（动词在前，主语在后，但中文翻译时一般不倒装）。

2. Here, only, so, hence, then 等词放在句首

Here ,only, so, hence, then 等词放在句首，这时主语往往放在动词的后面。

Here are some examples.

举例如下（这是句比较典型的句子，主语在后）：

Steel is a good conductor of electricity, and so **are** most metals.

钢是良导体，大多数金属也是良导体（动词在前，主语在后）。

Only after pressing the "CLASS BEGIN" **can** other button be controllable.

只有按下了"上课"按钮后，其他按钮才受控（动词在前，主语在后）。

3. 某些状语从句

如 no matter how, no matter what, however, whatever 等连词引导的让步状语从句，常把所修饰的宾语放在最前面。

Electrons are universal constituents of matter, their mass and electric charge being the same from whatever **element** they are obtained.

电子是物质的普遍成分，无论来自于哪种元素，电子的质量和电荷都是相同的。

As（though）引导的让步状语从句结构则为：表语+as（though）+主语+助动词，如：

Light as aluminum is, its strength is great.

铝虽然轻，但强度很高。

当 as 引导的从句中有助动词时，可以把动词放在前面。

Try as you may, you cannot be successful.

你可以去试试，但你不会成功的。

有时甚至让步状语从句不用连词，而用倒装句，这时谓语的一部分通常是 be，且用原形，放在句首，翻译时要注意。

Nearly all our clothes are made from fibers of one sort or another, **be** they derived from where.

我们所有的衣服几乎都是用某种纤维制成的，无论这些纤维来源于何处。

As 或 just as 引导的方式状语从句和 than 引导的比较状语从句一般用倒装句，句中的谓语往往用 do 代替主句中的实义动词。

We may conclude that air occupies space, just as **does** any other fluid.

我们可以得出结论，空气像任何流体一样也占据空间。

The engine is controlled by a microcontroller, as **are** the anti-lock brakes and so on.

发动机是用单片机控制的，防抱死制动（刹车）系统等也是用单片机控制的。
The molecules of the liquid have more energy than **do** the molecules of the solid.
液体中的分子比固体中的分子具有更多的能量。

13.4 Exercises

1. Put the Phrases into English（将下列词组译成英语）
(1) 工业自动化
(2) 复杂系统
(3) 车库入口
(4) 信号灯
(5) 产品包装
(6) 传输线
(7) 供水阀
(8) 实际系统
(9) （装配产品）流水线
(10) 安装发动机

2. Put the Phrases into Chinese (将下列词组译成中文)
(1) industrial machinery
(2) dealing with a simple system
(3) there is no space available
(4) solve the classic packaging problem
(5) tank full switch is off
(6) a significant change in rate
(7) minimize the motion of the valve
(8) a manufacturing process
(9) the modern consumer culture
(10) installation of both the hood and the wheels

3. Sentence Translation (将下列句子译成中文)
(1) When 100 cars park, a signal will turn on signalizing that a garage is full and notifying other drivers not to enter because there is no space available.
(2) This way lamp will emit light when a number of cars is greater than or equal to 100.
(3) The conveyor takes a box up to the limit switch, and a motor stops then.
(4) Using only digital signals, the PLC has two digital inputs from switches (tank empty and tank full).
(5) Assembly lines don't increase the speed of producing a single unit, but only increases rate

when there are a stream of units to be produced.

4. Translation (翻译)

Numerical control or numerically controlled (NC) machine tools are machines that are automatically operated by commands encoded on a digital medium. NC machines were first developed soon after World War II and made it possible for large quantities of the desired components to be very precisely and efficiently produced (machined) in a reliable repetitive manner.

13.5 课文参考译文

自动化或工业自动化是用控制系统如计算机或 PLC 去控制工业机器和工作过程，尽量减少人的工作（干预）。

在全球经济发展中，自动化发挥了日益增长的重要作用。工程师们一直在做这样的工作，结合自动化设备与数学及管理工具为迅速发展的各种人类活动和应用创建复杂的系统。这里是几个自动化的简单例子。

13.5.1 自动化停车场

这里讨论一个最多可以停 100 辆车的简单的自动化控制系统（图 13.1）。每当一辆车驶入，PLC 自动把在停车场内所有的车辆总数加一。每当一辆车开走，就把在停车场内所有的车辆总数减一。如果已停满 100 车，则信号灯亮表示停车场已停满，提醒外面的驾驶员不要再开入停车场，因为那儿已经没有空位。

来自停车场入口处的传感器信号设置 IR200.00 位，这个二进制位在程序中是执行以下两条指令的条件，第一条指令是把 CY 位清零（总是把这位先清零，因为后面的一些计算会改变它的值），另一条指令是把存储在字 HR00 中的车辆的总数加一并仍保存在 HR00 中。HR00 存储空间用来存储车辆的总数，因为这样就保存已停车辆的状态。在加法和减法指令中所用到的"#"符号定义用十进制数来对已停在停车场中的车辆数作加法或减法。因为条件 SR253.13 总是置 1，因此一定执行比较指令 CMP，这就意味着无论有车进入还是有车离开停车场，程序的每个循环中都要作比较运算。"停车场满"的信号灯是接在一个输出口 IR010.00，由地址为 SR255.06 的 EQ（相等）标志和地址为 SR255.05 的 GR（大于）标志控制的。这二位输入值相"或"的结果就是 IR010.00 的输出，用于控制信号灯。当车的数目大于或等于 100 时灯就亮了。在实际运作中车辆的数目可能大于 100 是因为有些不相信（停车场已满）的驾驶员可能会开进去看看是否还有空地（可以停车）。所以车的当前数目有可能从 100 增加到 101。当这个（不相信停车场已满的）驾驶员开车离开时，停车场的数目又回到 100。实际中很多停车场都有这种情况。

13.5.2 自动化包装线

产品包装线是工业自动化中最常见的自动化系统之一。可以在小型的机器中看到（例如包装食品等小颗粒的东西，也会在大型的系统中看到如包装药品等。这里以经典的包装问题（图 13.2）为例，只需要很少几个自动化的动作。需要用 CPM1A PLC 控制器的几个输入输出端。

按下"启动"健，就把标志位 Flag 1 置 1（程序段 1），Flag 1 置 1 是执行程序的条件（只有按下"停止"键才可以清零）。当启动后，带动箱子传输线的电动机开始工作，传输线带着箱子运动到极限开关（行程开关），则电动机停止（程序段 4）。行程开关实际上又是苹果传输线电动机的启动开关，当箱子到位后，带有苹果的传输线开始移动（程序段2），通过一个传感器对苹果计数，一个箱子装 10 个苹果，计数值到 10 计数器 CNT010 产生标志位，使得箱子传输线又开始带着箱子移动（程序段 3），当箱子传输线开始移动时，行程开关使计数器清零，为下次数苹果作准备。这个过程一直重复，除非按下"停止"按键使 Flag 1 清零，（程序就不再执行，停止工作）。图 13.3 给出了包装线信号的时间流程。

13.5.3　水箱水位控制

一个用来储水的水箱，其中的水要供应另一个系统，因此系统需要对水箱中的水位进行控制，有数字信号和模拟信号弹两种信号可用来控制。

如果仅用数字信号，PLC 需要两个开关量的数字信号输入（水箱空和水箱满）进行控制，PLC 输出一个数字信号用来打开和关闭水箱的供水阀。控制过程的梯形图如图 13.4 所示。

当水位掉到使得水箱空开关动作（动合触点闭合）时，PLC 就打开阀门让水流入。一旦水位升到足够高，水箱满开关动作（动断触点断开）时，PLC 关掉阀门不让水过多。

模拟控制系统（图 13.5）则可以用水压传感器或重量传感器和一个可调流量的水箱，调节流入水箱水流的阀门组成。

在这个系统中，为了避免不断来回调节损耗阀门，许多 PLC 动作的滞后性使得 PLC 有一个调节的死区。校正这个死区的一种技术是使阀门只在移动速度上做了一点改变，这样就可以把阀门的运动最小化，减小它的磨损。

一个实际的系统可以是两种方法（数字和模拟）的组合，用开关和简单的阀门来防止溢出，一个速率传感器和一个速率阀去优化供水的速率，防止水击作用（损坏器件）。因为还要考虑备用和维护措施等，实际的系统会复杂得多。

13.5.4　（装配）流水线

流水线是一个制造过程，在流水线中，按用一种优化的逻辑顺序不断加入零件组合成最终产品，生产速度比手工方式快得多。最有名的流水线，是福特电机公司在 1908～1915 年之间实现并进行生产的移动流水线。用流水线实现大批量生产被普遍认为是通过尽可能降低商品的成本造成现在消费文化的原因（催化剂）。

以汽车装配流水线为例，假设流水线的步骤为：安装发动机，发动机罩，车轮，（按这个次序，步骤中间可以有空隙），流水线上的汽车一次只能做这三步中的一步。汽车装上发动机后，就移动过去装发动机罩，把装发动机的位置让给下一辆汽车。接着第一辆汽车移去装轮子。第二辆汽车去装发动机罩，第三辆去装发动机。如果发动机安装要 20 分钟，发动机罩安装要 5 分钟，轮子安装要 10 分钟，则如果一次只能安装一辆汽车要完成三辆汽车的安装要 105 分钟。

另外，用流水线，因为有三个不同的安装站，发动机安装站，发动机罩安装站和轮子安装站，则三辆汽车总的安装时间是 75 分钟。因为安装发动机要 20 分钟，而完成发动机罩和轮子的安装共需 15 分钟，瓶颈是在发动机安装。因此，后面每隔 20 分钟就有一辆汽车从流水线上开出来。

流水线并不能增加生产一个产品的速度，但是当有一队产品要生产时流水线提高了生产速度。在上面的例子中，安装一辆汽车仍要 35 分钟，但汽车排队安装时只要 20 分钟就有一辆汽车从流水线开出来。

13.6 阅读材料参考译文

13.6.1 变频调速（驱动）系统

变频调速系统（VFD）是通过控制供给电动机的频率来控制交流（AC）电动机的旋转速度。变频率驱动是调速驱动的一种形式。变频调速也称为可调频率驱动（AFD）、变速驱动器、交流驱动器、微驱动器或逆变设备。因为输出电压随着频率变化而变化，有时也称为变压变频（VVVF）设备。

一个变频调速系统一般由交流电动机、变频控制器和操作界面组成（图 13.6）。

变频调速系统中所用的电动机一般是三相异步（感应）电动机，有时也用单相异步电动机，但三相电动机用得比较多。在有些场合用各种同步电动机有优势，但对大部分应用来说，异步电动机更适用，是更加经济的选择。一般选用主电路电压下按固定速度运行的电动机，但在标准电动机的设计上作一定的改进，使之工作可靠性更高，并有较好的变频调速性能。

变频驱动控制器是应用固态电子技术的功率变换设备，通常的设计是先把输入的交流电通过桥式整流转换成直流电（作为中间量），然后再把直流电用一个逆变开关电路转换成类似正弦的交流电源。整流电路通常采用二极管构成三相桥式整流电路，但有时也用可控整流电路（图 13.7）。

操作者用操作界面来启动或停止电动机，调节电动机的速度。附加的操作控制功能包括使电动机反转，在手动速度调节和根据外部处理控制信号自动控制之间切换。操作界面通常有图形界面显示，指示灯和仪表用来表示（系统）驱动运行的信息。

13.6.2 机器人及其控制

机器人（图 13.8）有以下的一些特征，虽然并不一定有全部以下特征：
（1）是非自然的，即是人造的。
（2）能够感知周围的环境，并处理其中的一些信息或与它们互动。
（3）有根据环境做出选择的能力，通常是用自动控制或编程的方法实现的。
（4）可以朝一个方向或多个方向转动或平动。
（5）有目的或间接目的的动作。

机器人的机械结构必须受控才能完成动作。机器人的控制包括感知、处理和动作三个阶段。传感器给出周围环境或机器人本身的信息，如机器人关节（连接处）的位置或者它的处理对象的位置。然后处理这个信息，计算出相应的信号送给执行器，执行器就使机械结构动作。

处理阶段可能有不同的复杂度。首先从含有噪声的传感器信号中估算出所需的参数，如机械手（夹子）的位置。从这些参数中推断出要执行的任务（如机械手在某个方向移动），根据控制理论把这些任务转换成驱动执行器的命令。

如果控制时间比较长或任务比较复杂，机器人需要建立一个"辨识"模型并进行推理，辨识模型用于表示机器人，外部世界以及他们的相互作用。图案辨识和计算机视频可用来追踪目标，地图技术可用于建立世界（外部环境）地图。最后，动作策划和其他人工智能技术可用于计算出如何动作。

控制系统的自动化程度也各不相同，在触觉或远程操作设备中通过直接相互作用，人们几乎完全控制机器人的行动。在辅助操作模式中操作者给出中高级难度的任务命令，机器人自动算出如何完成这些任务。自动化机器人可以在没有人类干预情况下工作很长时间。较高程度的自动化并不需要更复杂的辨识能力，例如，在工厂流水线上的机器人可以完全自动化，但以一个固定的模式工作。

13.6.3　过程控制系统

过程控制系统一般是由一组电子器件和设备组成，起到使生产过程稳定，精确和消除有害的暂态效应的作用。

过程控制系统可能有不同的形式和执行方式，从输送能量到加工产品等。随着技术快速发展，许多复杂的操作任务都可以通过连接可编程控制器，可能还有中央计算机来完成了。除了可以直接连接仪器，如操作台、电动机、传感器、开关、阀门外，这些仪器（与可编程控制器连接后）有可能互相之间传递信息，因此，可以在很高水平上进一步开发和过程合作，同时使得这样一个过程控制系统有较大的可调节性。过程控制系统的每一个无件，无论大小，都起着重要的作用。例如，没有传感器，PLC 就不能准确地知道过程进行到什么地步。在自动化系统中，PLC 通常是一个过程控制系统的中心部件，通过执行储存在存储器中的程序，PLC 通过输入器件输入的信号不断监控着系统的工作状态，在程序的逻辑运算的基础上，PLC 决定下一步采取什么动作并输出信号，通过外部的仪器执行完成所需的动作。要执行更复杂的过程，可能要通过中央计算机连接更多的 PLC。图 13.9 所示为一个实际系统的框图。

Unit 14 Electronic Design Automation

14.1 Text

Electronic design and applications is a vital area of electrical engineering, encompassing the experimentation, design, modeling, simulation and analysis of single devices or circuits as well as complete signal processing systems. Electronic circuit design has a lot of work to do, and now the computer can be used to help complete the electronic circuit design and debugging.

14.1.1 What is EDA

Electronic Design Automation or Electronic Design Application (EDA), is the category of tools for designing and producing electronic systems (Fig 14.1 are some interfaces of EDA) ranging from printed circuit boards (PCBs) to integrated circuits. EDA involves a diverse set of software algorithms and applications that are required for the design of complex next generation semiconductor and electronics products. This is sometimes referred to as ECAD (electronic computer-aided design) or just CAD.

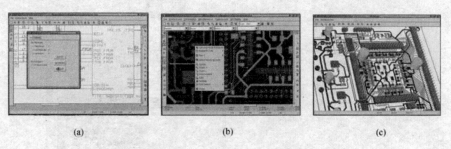

Fig 14.1 some interface of EDA
(a) Circuit design; (b) PCB layout; (c) 3D View

While early EDA focused on digital circuitry, many new tools incorporate analog design, and mixed systems. This is happening because there is now a trend to place entire electronic systems on a single chip.

Current digital flows are extremely modular. The front ends produce standardized design descriptions that compile into invocations of "cells", without regard to the cell technology. Cells implement logic or other electronic functions using a particular integrated circuit technology. Fabricators generally provide libraries of components for their production processes, with simulation models that fit standard simulation tools.

Analog EDA tools are much less modular, since many more functions are required, they interact more strongly, and the components are (in general) less ideal.

The circuit design is the very first step of actual design of an electronic circuit. Typically sketches are drawn on paper, and then entered into a computer using a schematic editor. Therefore schematic entry is said to be a front-end operation of several others in the design flow.

Despite the complexity of modern components—huge ball grid arrays and tiny passive components—schematic capture is easier today than it has been for many years. CAD software is easier to use and is available in full-featured expensive packages, very capable mid-range packages that sometimes have free versions and completely free versions that are either open source or directly linked to a printed circuit board fabrication company.

14.1.2 What is EWB?

Electronics Workbench is a design tool that provides you with all the components and instruments necessary to create board-level designs. Fig 14.2 is the interface of EWB. It has complete mixed analog and digital simulation and graphical waveform analysis, allowing you to design your circuit and then analyze it using different simulated instruments and analysis options. It is fully integrated and interactive, thus you can change your circuits quickly, allowing fast and repeated what-if analysis.

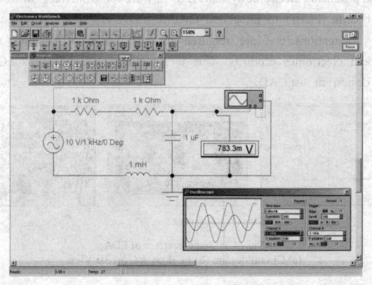

Fig 14.2 the interface of EWB

1. Circuit Simulation Mechanism

After you create a circuit schematic and turn on the power or click the simulate button, the solution of the circuit and generation of the data you see on instruments such as the oscilloscope is the role of the simulator. More specifically, the simulator is the part of Electronics Workbench that calculates a numerical solution to a mathematical representation of a circuit you created.

For this calculation to occur, each component in a circuit is represented by a mathematical model. Mathematical models link the schematic in the circuit window with the mathematical representation for simulation. The accuracy of the component models determines the degree to which simulation results match real-world circuit performance.

The mathematical representation of a circuit is a set of simultaneous, nonlinear differential equations. The main task of the simulator is to solve these equations numerically. A SPICE-based simulator transforms the nonlinear differential equations into a set of nonlinear algebraic equations. These equations are further linearized using the modified Newton-Raphson method. The resulting set of linear algebraic equations is efficiently solved using the sparse matrix processing LU factorization method.

2. Four Stages of Circuit Simulation

The simulator in Electronics Workbench, like other general-purpose simulators, has four main stages: input, setup, analysis and output.

(1) At the input stage, after you have built schematic, assigned values and chosen an analysis, the simulator reads information about your circuit.

(2) At the setup stage, the simulator constructs and checks a set of data structures that contain a complete description of your circuit.

(3) At the analysis stage, the circuit analysis specified in the input stage is performed. This stage occupies most of CPU execution time and actually is the core of circuit simulation. The analysis stage formulates and solves circuit equations for the specified analyses and provides all the data for direct output or post-processing.

(4) At the output stage, you view the simulation results. You can view results on instruments such as the oscilloscope, or on graphs that appear when you run an analysis from the Analysis menu or when you choose Analysis/Display Graphs.

14.1.3 Design with Protel 99

Several electronic design applications exist to both create schematics of a circuit and transfer them to a working PCB layout. Here we discuss the design of a circuit using Protel 99SE (Fig 14.3).

(1) Start Protel and select **File/New**. Use a MS Access Database and give it a name – this will create a Design Database file (extension.DDB) that contains all the parts of your design.

(2) The Database has its own internal File System–go into the Documents Folder, select **File/New** and create a new Schematic Document that will describe our circuit. This consists of placing symbols that represent the individual components in the design and connecting their individual connections, or nodes, to one another.

(3) We are now ready to check our design to make sure there are no errors. This can be done using what called an Electrical Rule Check (ERC). It can be found under **Tools/ERC**.

(4) We are now ready to actually create a PCB from this schematic. There is no information about physical data in a schematic–orientation, spacing. The only data that will be sent to the PCB document besides a list of what nodes are connected to each other (a netlist) are the number of pins on each part, and the Footprint for each part.

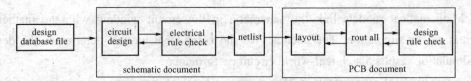

Fig 14.3 the process of design with Protel 99 SE

(5) We can first layout the components. Clicking a component and dragging the mouse will allow you to move the component. Pressing the Space Bar while doing this causes it to rotate. You should try to place the components reasonably close together (save surface area), and in a manner that ensures that the connections cross each other as little as possible. This will make it easier for Protel to route the design.

(6) We now need to tell Protel's AutoRouter to work only in the Top Layer and select **AutoRoute/All**, and hit **Route All**. The Board should Route.

(7) Now we can run a Design Rule Check (DRC), which will ensure that we have no short circuits, that everything is routed, etc. It can be accessed from **Tools/Design Rule Check**.

The PCB design is now complete. All that remains is to transfer the PCB out of Protel in a format that can be understood by the PCB manufacture.

 Review

(1) Electronic design application (EDA) is the category of tools for designing and producing electronic systems ranging from printed circuit boards (PCBs) to integrated circuits.
(2) Electronics Workbench is a design tool that provides you with all the components and instruments necessary to create board-level designs.
(3) Protel 99SE can both create schematics of a circuit and transfer them to a working PCB layout.

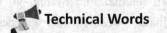 Notes to the text

[1] modified Newton-Raphson method 改进的牛顿·拉夫逊算法。

Technical Words

algebraic [ˌældʒiˈbreiik] *a.* 代数的
compile [kəmˈpail] *vt.* 编译，编辑，汇编
ensure [inˈʃɔ:] *vt.* 保证，担保，使安全 *v.* 确保，确保，保证
fabricator [ˈfæbriˌketə] *n.* 捏造者，制作者，假造者，杜撰者
incorporate [inˈkɔ:pəreit] *adj.* 合并的，结社的 *vt.* 合并，使组成公司 *vi.* 合并
interactive [intərˈæktiv] *adj.* 交互式的；相互作用的
invocation [ˌinvə(ʊ)ˈkeiʃ(ə)n] *n.* 祈祷，（计）调用

layout ['leiaut] n. 电路布线图，配线，设计图案，（工厂等的）布局图
orientation [ˌɔːriən'teiʃ(ə)n] n. 方向，方位，定位
oscilloscope [ə'siləskəup] n. 示波器
simulation [ˌsimju'leiʃən] n. 仿真，假装，模拟
simulator ['simjuleitə] n. 仿真器，模拟者

Technical Phrases

printed circuit board	印刷电路板，简称 PCB
cell technology	模块技术（组成一个小单元电路）
schematic editor	电路编辑器
nonlinear differential equation	非线性差分方程
algebraic equation	代数方程
netlist	网络列表，电路器件连线表
Schematic Document	（设计）原理图文件
Electrical Rule Check	电气规则检查

14.2 Reading materials

14.2.1 Power systems CAD

Power systems CAD refers to computer-aided design (CAD) software tools that are used to design and simulate complex electrical power systems. Electrical power systems CAD tools are used by electrical power systems engineers, a distinct discipline of electrical engineering.

According to the Institute of Electrical and Electronics Engineers (IEEE), there are 21 000 power systems engineers worldwide focused on improving electrical grids, eliminating blackouts, and reducing electrical accidents. Such engineering expertise is instrumental to preserving the critical power needs of modern digital society, e.g. transportation, communications, computing, etc.

Power systems CAD tools have following virtue:

(1) Providing a design foundation that allows power systems to be created quickly.

(2) Enabling design engineers to test the safety and integrity of their design concepts.

(3) Allowing design engineers to create a repository if proven design elements that can be reused in future projects.

Thus power systems CAD software products allow organizations to develop higher-quality power systems designs. The electrical power systems CAD process, frequently called power systems "modeling", typically consists of two distinct stages:

(1) The design stage, in which an electric systems model is created.

(2) The simulation or analysis stage, in which software simulation programs are used to test the integrity of the design; these simulation programs test how the model would behave in real-world operation by checking for specific types of design or operational problems.

It is important to note that this is an iterative process, in which simulation results will suggest ways that the design should be modified to increase safety, reliability, and serviceability. At the conclusion of the design effort, organizations will enjoy a far higher degree of confidence in the integrity of their power systems infrastructure than with manually-drawn schematics.

There are a wide range of electrical engineering tests that can be performed on a power systems CAD model. Here are some examples:

(1) Short Circuit Analysis

(2) Protective Device Coordination

(3) Power System Reliability

(4) Electromagnetic Transient Analysis

(5) Transmission Line Parameters

14.2.2 Simulation

Simulation is the imitation of some real thing, state of affairs, or process. The act of simulating something generally entails representing certain key characteristics or behaviors of a selected physical or abstract system.

Simulation is used in many contexts, including the modeling of natural systems or human systems in order to gain insight into their functioning. Other contexts include simulation of technology for performance optimization, safety engineering, testing, training and education. Simulation can be used to show the eventual real effects of alternative conditions and courses of action.

A computer simulation is an attempt to model a real-life or hypothetical situation on a computer so that it can be studied to see how the system works. By changing variables, predictions may be made about the behavior of the system.

Computer simulation has become a useful part of modeling many natural systems in physics, chemistry and biology, and human systems in economics and social science as well as in engineering to gain insight into the operation of those systems.

For an example, A flight simulator is used to train pilots on the ground. It permits a pilot to crash his simulated "aircraft" without being hurt. Flight simulators are often used to train pilots to operate aircraft in extremely hazardous situations, such as landings with no engines, or complete electrical or hydraulic failures. The most advanced simulators have high-fidelity visual systems and hydraulic motion systems. The simulator is normally cheaper to operate than a real trainer aircraft.

14.2.3 SPICE

SPICE (Simulation Program with Integrated Circuit Emphasis) is a general purpose analog electronic circuit simulator (Fig 14.4). It is a powerful program that is used in IC and board-level design to check the integrity of circuit designs and to predict circuit behavior.

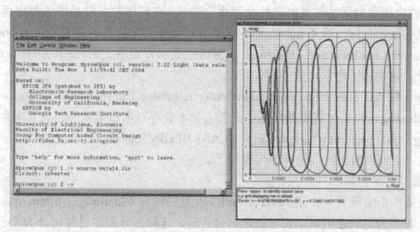

Fig 14.4   interface of SPICE

Integrated circuits, unlike board-level designs composed of discrete parts, are impossible to breadboard before manufacture. Further, the high costs of photolithographic masks and other manufacturing prerequisites make it essential to design the circuit to be as close to perfect as possible before the integrated circuit is first built. Simulating the circuit with SPICE is the industry-standard way to verify circuit operation at the transistor level before committing to manufacturing an integrated circuit.

Board-level designs can often be breadboarded, but designers may want more information about the circuit than is available from a single mock-up. For instance, performance is affected by component manufacturing tolerances and it is helpful for designers to simulate with SPICE to predict the effect of variations of those values. Even with a breadboard, some aspects may not be accurate compared to the final printed wiring board, such as parasitic resistances and capacitances. In these cases it is common to perform Monte Carlo simulations using SPICE, a task which is impractical using calculations by hand.

SPICE inspired and served as a basis for many other circuit simulation programs, in academia, in industry, and in commercial products. SPICE became popular because it contained the analyses and models needed to design integrated circuits of the time, and was robust enough and fast enough to be practical to use.

14.3   Knowledge about translation（否定的表示）

英语与汉语在表达否定概念时所使用的语言手段有很大差别，因此，在理解与翻译时要特别加以注意。科技英语中常用否定语气的结构可分为全部否定、部分否定、双重否定和意义上的否定。

1. 全部否定

用 not 否定谓语动词是常见的一种全部否定形式。在翻译时一般都译成否定谓语，这与汉语的否定结构基本相同，除了 not 以外，其他表示全部否定意义的词有 no、nobody、none、nowhere、never、neither、nor、nothing，不管这些表示否定意义的词在句中作主语、宾语还是其他成分，这类句子通常译成否定句。

Nobody who has ever seen good quality color television can ever be completely happy with black and white again. 见过高质量彩色电视的人是不会再对黑白电视感到完全满意的。

Nothing would perform when pressing "CLASS BEGIN" again.

再次按"上课"按钮时，没有任何作用了。

2. 部分否定

英语中某些不定代词（如 all，every，both）及某些副词（如 always，often，quite，entirely，altogether）等与否定词连用时，表示的是部分否定。这种部分否定通常可译为"不全是""不都是""不常""未必都""并非完全"等。

Not all substances are conductors.

并非所有的物质都是导体。

The electrons within a conductor are not entirely free to move but are restrained by the attraction of the atoms among which they must move.

导体中的电子运动并不是完全自由的。电子必须在原子之间运动，从而要受到这些原子引力的束缚。

A programmer can avoid the use of assembly code in all but most demanding situations.

一个编程人员除了必须（要用汇编语言）情况下可以避免使用汇编语言（编程）（部分否定）。

3. 双重否定

双重否定结构通常是由 no，not，never，nothing 等词与含有否定意义的词连用而构成的。这种结构形式上是否定，实质上是肯定，语气较强。翻译时可译为双重否定，有时也可译成肯定句。

You can do nothing without energy.

没有能量，你就什么也做不成。

In fact, there is hardly any sphere of life where electricity may not find useful application.

事实上，几乎任何一个生活领域都要用到电。

4. 意义上的否定

英语中有些词和词组在意义上表示否定。如：little（几乎没有），few（几乎没有），seldom（极少），scarcely（几乎不），hardly（很难，几乎不），too…to（太……以致不……），rather than（而不）fail to（不成功……，未能……）等。翻译时要译出否定的意义。

Metals, generally, offer little resistance and are good conductors.

通常金属几乎没有电阻，因而是良导体。

Glass conducts so little current that it is hardly measurable.

玻璃几乎不导电，因此很难测量其中的电流。

It may be easier to apply a force by pushing down rather than by pulling up.

向下推容易用力，向上举不容易用力。

Older designs employed audio interstage and output transformers but the cost and size of these parts has made them all but disappear.

比较早的设计是用内部多级音频放大器和输出变压器，但这些部件的成本和体积因数使这些设计被淘汰。

5. 否定转移

因英语与汉语在表达否定要领时所使用的词汇手段与语法手段都有很大的差别，所以在翻译时常用到两种转移，一是语法的转移，即否定主语或宾语转移成否定谓语，否定谓语转移成否定状语等；二是内容上的否定转移，即英语中的否定形式译成汉语时可用肯定形式，反之亦然。

No smaller quantity of electricity than the electron has ever been discovered.
从来没有发现过比电子电荷更小的电量（由否定主语转移为否定谓语）。

Electric current cannot flow easily in some substances.
电流不能顺利地在某些物质中流动（从逻辑上判断是否定状语 easily）。

Its importance cannot be stressed too much.
它的重要性怎么强调也不过分（不要误译为：它的重要性不要强调的太过分）。

We have seen that the beta rays are nothing but a stream of electrons.
我们已经知道，β 射线只不过是一种电子流。

14.4 Exercises

1. Put the Phrases into English (将下列词组译成英语)
(1) 电子设计软件
(2) 标准化设计
(3) 器件库
(4) 仿真模型
(5) 印刷电路板
(6) 电路性能
(7) 器件模型
(8) 分析阶段
(9) 电路原理图
(10) 电气规则检查

2. Put the Phrases into Chinese (将下列词组译成中文)
(1) integrated circuits
(2) electronic computer-aided design
(3) graphical waveform analysis

(4) using different simulated instruments
(5) repeated what-if analysis
(6) click the simulate button
(7) the accuracy of the component model
(8) real-world circuit performance
(9) nonlinear differential equations
(10) check our design

3. Sentence Translation (将下列句子译成中文)

(1) Fabricators generally provide libraries of components for their production processes, with simulation models that fit standard simulation tools.

(2) Typically sketches are drawn on paper, and then entered into a computer using a schematic editor.

(3) It has complete mixed analog and digital simulation and graphical waveform analysis, allowing you to design your circuit and then analyze it using different simulated instruments and analysis options.

(4) Several electronic design applications exist to both create schematics of a circuit and transfer them to a working PCB layout.

(5) You should try to place the components reasonably close together (save surface area), and in a manner that ensures that the connections cross each other as little as possible.

4. Translation (翻译)

Electronic memory comes in a variety of forms to serve a variety of purposes. Flash memory is used for easy and fast information storage in such devices as digital cameras home video and game consoles. It is used more as a hard drive than as RAM. In fact, Flash memory is considered a solid state storage device. Solid state means that there are no moving parts —— everything is electronic instead of mechanical.

14.5　课文参考译文

电气工程的一个重要领域是电子电路设计和应用，包括单个器件电路和完整信号处理系统的实验、设计、建模、仿真和分析。电子电路设计有很多工作要做，现在计算机可以用来帮助完成电子电路设计和调试。

14.5.1　EDA 是什么？

电子设计自动化或电子设计应用软件（EDA）是用来设计和制作从印刷电路板（PCB 板）到集成电路等各种电子系统的工具（图 14.1 为一些电子设计软件的界面），EDA 包括各种软件算法和应用程序，可以用来设计复杂的新一代半导体电子产品，有时还称为电子计算机辅助设计（ECAD）或计算机辅助设计（CAD）。

虽然早期的 EDA 的重点是设计数字电路，但许多新的（设计）工具综合了模拟电路设计和混合系统的设计。因为现在有一种发展趋势是把整个电子系统集成在单一芯片上。

当前数字（电路设计）流程是完全模块化的。前端设计产生标准的设计描述，提出想要的单元（性能）而不必考虑单元的技术。单元设计则用一些特殊的集成电路技术实现其逻辑功能或其他电路功能。厂商一般为他们的生产流程提供器件库，并提供适合标准仿真工具的器件仿真模型（即可以对他们的器件组成的电路进行仿真）。

模拟电路 EDA 软件的模块化程度比较低，因为对模拟电路的仿真需要更多的功能，电路之间的相互影响也更多，一般来说器件也不是十分理想化。

（原理）电路设计是实际电子电路设计的第一步。典型的（原理）电路图是画在纸上，然后用电路编辑软件输入到计算机中，所以在设计流程中原理图的输入是其他几个过程的最前端工作。

尽管现在的元器件越来越复杂——从微小的无源器件到巨大的球形网络分布，但原理图的设计却比多年前容易了，计算机辅助设计软件越来越好用，有比较贵的软件包，含有全部设计功能的，有很多功能也比较强的（中等）软件包，有时有免费版本的中等软件包，还有完全免费的版本，其源是开放式的，或者是直接连接到一个印刷电路板制作公司的。[公司提供免费设计软件，并（收费）帮你加工制作印刷电路板。]

14.5.2 EWB 是什么（软件）？

电子工作台（EWB）是一个设计工具，它提供所有的元件和必要的仪器用以进行电子线路板的设计。图 14.2 是 EWB 软件的界面。它可以完成模拟信号和数字信号的混合电路仿真和波形图分析。你可以用它设计电路并用不同的仿真仪器和分析选项进行分析。软件是完成集成的，有很好的交互性，因此你可以很快地修改电路，进行快速和反复的假设分析。

1. 电路的仿真

创建了一个电路图后，按下电源开关或仿真按钮，电路求解并产生数据，在仪器如示波器上显示出结果，这就是仿真器的作用。很显然，仿真器是 EWB 的一部分，用来对创建的电路的数学表达式进行一系列求解计算。

为了得出计算结果，电路中的每个元件都用一个数学模型表示，数学模型把电路窗口中的电路与仿真用的数学表达式相连接，元件模型的精度决定了仿真结果与实际电路性能相符合的程度。

一个电路的数学表达式是一组同时的，非线性的差分方程，仿真器的主要任务是求出这些方程的数值解。基于 SPICE（软件名）仿真器把非线性差分方程组变换成一组非线性的代数方程，这些方程用改进的牛顿—拉夫逊方法进一步线性化。导出一组线性代数方程后用稀疏矩阵 LU 因子分解方法进行求解。

2. 电路仿真的四个步骤

EWB 仿真器，就像其他通用仿真器一样，有四个主要的步骤：输入、设置、分析和输出。

在输入阶段，建立一个电路，指定参数值，选取一种分析方法，仿真器读入电路的信息。

在建立阶段，仿真器构造并检查对电路作完整描述的一组数据结构。

在分析阶段，进行在输入阶段指定的电路分析，这个阶段占用电脑最多的执行时间，实际上是电路仿真的核心。分析阶段对指定的分析建立方程并对电路方程组求解，给出可以直接输出或预处理的全部数据。

在输出阶段，就可以看到仿真结果了。可以在仪器如示波器上看到结果，如果你在分析菜单上选取一个分析时或者选择分析/显示图片时会用图片显示结果。

14.5.3　用 Protel 99 进行电路设计

有些电子设计软件是可以创建电路原理图并把原理图转换成有效的印刷电路图。这里我们讨论如何用 Protel 99 进行电路设计（图 14.3）。

（1）启动 Protel 并选择"文件/新建"，新建一个 MS 数据库类型的文件并命名。这样就创建了一个设计数据库文件（扩展名为.DDB），你设计的所有内容都会存在里面。

（2）数据库有它自身内部的文件系统——进入文件目录，选择"文件/新建"，建立一个新的用来表示我们电路的原理图文件。这个文件可以放置设计中表示单一元件的符号，这些元件相互的连接，或者与节点的连接。

（3）现在我们要检查我们设计，保证没有错误，这可以用电气规则检查（ERC）来进行，点击"工具/ERC"菜单就可以了。

（4）现在要根据原理图来创建印刷电路板了，但是原理图中除了输出表示哪些节点是相互连接的表（网络表）外，输送给印刷电路板文件的数据只是每个器件的引脚数和每个器件的引脚点。在原理图中并没有关于器件物理数据的信息，如器件的大小和器件摆放的方向。

（5）先布置元件，点击元件并拖动鼠标可以移动元件，按空格键可以使元件旋转。要尽量合理放置元件，使它们比较紧凑（可以节省表面积）并使得连接线的相互交叉尽可能少一些。这样 Protel 布线时比较容易。

（6）现在把 Protel 的自动布线设置成只在顶层布线，选择"自动布线/全部"并点击"全部布线"，印刷板就布好线了。

（7）接着可以运行设计规则检查（DRC），它会检查电路中有没有短路，是不是布好全部线了等，点击"工具/设计规则检查"。

（8）印刷电路板已完成了，接下来要做的是把 PCB 从 Protel 中导出，形成一个可以被 PCD 制造商用的文件（格式）（可以去控制机器进行加工了）。

14.6　阅读材料参考译文

14.6.1　电力系统 CAD

电力系统 CAD 是指用于设计和仿真复杂的电力系统的计算机辅助设计软件（工具）。电气工程的一个（独特的）子学科电力工程的工程师们使用电力系统 CAD 软件。

根据电气电子工程师学会（IEEE）统计，有遍布全世界的 21000 名电力工程师在做电网改造，消灭突然断电事故和减小电气事故方面的工作。电力工程专业保护现代数字社会的关键电力需求，如交通、通信、计算机等。

电力系统 CAD 软件（工具）有以下优点：

（1）提供一个设计平台，可以很快地创建电力系统（模型）。

（2）使设计工程师可以对他们的设计理念进行安全性和完整性进行测试。

（3）这软件允许设计工程师们可以创建一个设计元素库，如果是经过验证的设计元素，可以在未来的项目中重复利用。

因此电力系统 CAD 软件产品可以让一些设计组织开发出较高质量的电力系统设计（方案）。电力系统 CAD 设计过程，通常称为电力系统建模，由两个阶段组成：

（1）设计阶段，创建一个电力系统模型。

（2）仿真或分析阶段，用软件仿真程序测度设计的完整性，这些仿真程序通过检查指定类型的设计或运行问题（假设出现故障时）等方法测试（电力系统）模型在实际运行中的性能。

要注意这是一个反复（设计调试）的过程，其中仿真结果会对提出对设计要作那些改进可以增加安全性、可靠性和适用性。作为设计的结论，设计组织更相信他们电力系统基础设施的完整性，而不是手画的电路原理图。

在电力系统 CAD 模型上可以进行的电力工程测试的种类很多，例如：

（1）短路分析。

（2）保护设备配合。

（3）电力系统可靠性。

（4）电磁暂态效应分析。

（5）输电线参数。

14.6.2 仿真

仿真是对一些真实事情、事件的状态或过程的模仿。仿真某些事一般是表现出选定的物理量或抽象系统的一些关键特征或性能。

在很多领域都可以用到仿真，包括为了深入了解自然系统或人类系统的功能，可以对它们建模仿真。还包括为优化性能、工程安全性、测试、训练和教育所做的技术仿真。仿真也可以用来显示当条件和过程改变时系统的最终实际效果。

计算机仿真是在计算机上建立一个实际生活或假设情况的模型，从而可以对系统是如何工作进行研究。通过改变变量，预计系统的性能会发生什么变化。

计算机仿真已成为非常有用的一部分，应用于模仿物理、化学、生物等许多自然系统和经济和社会科学中的人类系统及工程系统中，深入了解这些系统运行情况。

例如，一个飞行模拟器可用来在地面上训练飞行员，它可以让飞行员撞他的模拟"飞机"而不会受伤。飞行模拟器通常用来训练飞行员在极端危险的条件下驾驶飞机，如在发动机不工作，电气系统完全损坏或液压故障条件下着陆。最高级的模拟器有高度逼真的视频系统和液压运动系统。模拟器一般比真的训练飞机便宜。

14.6.3 集成电路仿真软件

SPICE（集成电路仿真软件）是一个通用模拟电路仿真器（图 14.4）。这是一个功能强大的程序，可以用在集成电路和印刷电路板设计中，可以检查电路设计的完整性并指出电路的性能。

与分立元件组成的电路板（级）的设计不同，集成电路在制造出来之前是不可能用面包板（进行设计调试的）。再进一步考虑到集成电路蚀刻封装和其他制作的高成本，就要求集成电路设计的基本点是集成电路在制作出来之前就尽可能地接近完美。用 SPICE 对电路模拟是工业中标准方法，用来在制造一个集成电路之前验证在晶体管级的电路性能。

电路板级的设计通常可以用面包板，但设计者可能想要获得更多的电路信息，那不是一个模型可以提供的。例如，电路性能还会受到器件制造精度的影响，这时用 SPICE 模拟可以对这些值的变化时可能对电路所产生的影响进行分析。即使用面包板（进行设计），与最终印刷线路板相比，有些方面例如寄生电阻和电容是不够精确的。在这些情况中，通常用 SPICE 进行蒙特卡罗法（随机抽样法）模拟，这是手工计算不可能做到的任务。

SPICE 在学术界、工业和商业产品中是权威性的，用作许多其他电路模拟程序的基础。SPICE 含有设计集成时域电路所需的分析和建模（功能），且在实际应用中功能强，速度快，因此应用很广。

Unit 15　User Manual

15.1　Text

As the modern education developing, various multimedia devices have been brought into many classrooms. For example: multimedia computer, VCR, VCD, projector, video presentation platform, electrical curtain and amplifier etc. It's not convenient for the teacher to manipulate all these above. The Education Control System is developed to resolve the problem. By controlling the multimedia devices on a computer or the control panel, the course of control becomes simple.

This manual is applied to PC-950/ PC-3900 Education Control System.

15.1.1　Safety Operation Guide

In order to guarantee the reliable operation of the equipments and safety of the staff, please abide by the following proceeding in installation, using and maintenance:

(1) When installing the equipment, make sure wire grounding in power cable is fine, DO NOT use double-legged socket, and ensure the voltage of input power supply consistent with which marked on the host.

(2) There is 110V/220V AC high voltage components inside, please DO NOT open the casing without permission, in case of electric shock.

(3) DO NOT put the system equipment in the place too cold or too hot.

(4) As the power generating heat when running, the working environment should be maintained fine ventilation, in case of damage caused by overheat.

(5) Please cut off the general power switch in humid weather or left unused for long time.

(6) Before following operation, ensure that the alternating current wire is pull out of the power supply:

① Take off or reship any components of the equipment.

② Take off or rejoin any pin or other link of the equipment.

(7) As to non-professional or without permission, please DO NOT try to open the casing of the equipment, DO NOT repair it on your own, in case of accident or increasing the damage of the equipment.

(8) DO NOT splash any chemistry substance or liquid in the equipment or around.

15.1.2 PC-950 Front Panel Introduction

The front Panel of PC-950 Education Control System is shown in Fig 15.1.

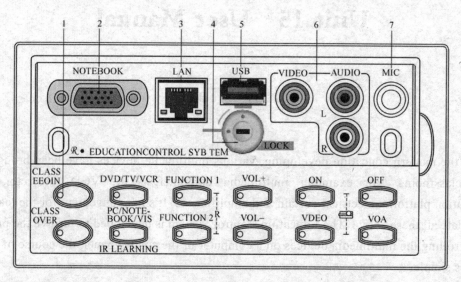

Fig 15.1　PC-950 front panel

1. Control button

(1) Begin or Finish Class

① Press "CLASS BEGIN", electrical screen moves down and projector power on. 7 seconds later the projector start working. Nothing would perform when pressing "CLASS BEGIN" again.

② Press "CLASS OVER", the system carry out finish class order at a fixed time. When perform the order, the electrical screen moves up and the projector will be cut off through RS232 code. At this moment, only "CLASS BEGIN" is controllable.

Notice：Only after pressing the "CLASS BEGIN" can other button be controllable.

(2) Projective Video Selection

① Press "DVD/TV/VCR" in order to transmit AV signal from DVD to projector, and transmit the RS232 code (the "DVD/TV/VCR" indicator light turns red when switch to DVD); Press for the second time, it will transmit AV signal from TV to projector(the "DVD/TV/VCR" indicator light turns green when switch to TV); Press for the third time, it will transmit AV signal from VCR to projector (the "DVD/TV/VCR" indicator light turns orange when switch to VCR).

② Press "PC/NOTEBOOK/VIS" in order to transmit VGA and audio signals from computer to projector, and send the RS232 code (the "PC/NOTEBOOK/VIS" indicator light turns red when switch to PC); Press for the second time, it will transmit VGA and audio signals from laptop computer to Education Control System CREATOR projector (the "PC/NOTEBOOK/VIS" indicator light turns green when switch to NOTEBOOK); Press for the third time, it will transmit VGA signal from digital platform to projector (the "PC/NOTEBOOK/VIS" indicator light turns orange when switch to VIS).

(3) **Volume Control**

① VOL+ output volume up

② VOL- output volume down

(4) **Projector Control Button**

① ON——Send IR and RS232 codes to open the projector when press "ON".

② OFF——Send IR and RS232 codes to close the projector when press "OFF".

③ VIDEO——Send IR and RS232 codes of projector video when press "VIDEO".

④ VGA——Send IR and RS232 codes of computer signal when press "VGA".

(5) **IR Learning**

The buttons include "FUNCTION1、FUNCTION2、ON、OFF、VIDEO、VGA"

Steps：

① Press "IR LEARNING" for 2 seconds, it will turn to IR Learning mode and other buttons become uncontrollable. The "IR LEARNING" will light for 10 seconds, and then it light off if not receives any IR code.

② Point the remote controller directly to IR receiving window and press corresponding buttons, then 6 IR learning buttons blink after receiving the IR code. It will quit learning mode 10 seconds later when not receives any IR code.

③ Press the learning button within 10 seconds in order to make IR learning successful. Then it lights off after blinking 3 times.

2. Portable Computer VGA Signal Input Port

Connect to portable computer VGA port, adopting standard 15HDF interface.

3. LAN

Directly connect to the Ethernet port on rear panel.

4. Electric lock

Only a special key can open this lock. It equals to the function of "CLASS BEGIN" and "CLASS OVER". No button is controllable when locked.

5. USB

Directly connect to the USB port on rear panel.

6. AV Signal Input

Input the AV signal from external devices to the system. By default setting, the AV signal transmits to amplifier or projector output.

7. Microphone Port

Connect to microphone

15.1.3 The Software Function

The main interface of education control system software is shown in Fig 15.2.

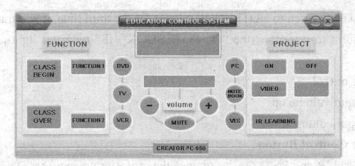

Fig 15.2　the main interface of education control system software

1. CLASS BEGIN

Left click mouse to begin class. Right click mouse to open following window (Fig 15.3), and you can set whether allow "Projector Action" and "Projector Screen". Tick what you need and click "Download" to confirm.

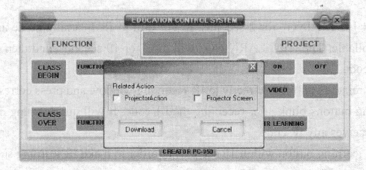

Fig 15.3　class begin linkage setup window

After choosing "Projector Action" and "Projector Screen", it will transmit RS232 code and IR code to start both projector and projector screen.

2. CLASS OVER

Left click mouse to finish class. Right click mouse to open following window (Fig 15.4), and you can set whether allow "Projector Action" and "Projector Screen". Tick what you need and click "Download" to confirm.

Fig 15.4　class begin linkage setting window

You can also set the delaying time.
Setting range Class Over: 0~60min
Setting range Projector: 2~6min
Then confirm it.
After choosing "Projector Action" and "Projector Screen", it will transmit RS232 code and IR code to close both projector and projector screen.

Review

(1) User manuals usually have "Safety Operation Guide".
(2) User manuals introduce how to connect and operate in detail with the help of many graphics.
(3) Nothing would perform when pressing "CLASS BEGIN" again.

Technical Words

blink [bliŋk] v. 眨眼，闪亮，闪烁
casing ['keisiŋ] n. 包装，保护性的外套
guarantee [gær(ə)n'ti:] n. 保证，保证书，担保，抵押品   vt. 保证，担保
manipulate [mə'nipjuleit] vt. (熟练地) 操作，使用 (机器等)，操纵 (人或市价、市场)，利用，应付，假造   vt. (熟练地) 操作，巧妙地处理
manual ['mænju(ə)l] n. 手册，指南   adj. 手的，手动的
projector [prə'dʒektə] n. 放映机，投影机，探照灯，设计者
reship [ˌri:'ʃip] vt. 把……再装上船   vi. 再上船，重装，转载
splash [splæʃ] n. 溅，飞溅，斑点   v. 溅，泼，溅湿
staff [stɑ:f] n. 全体职员   vt. 供给人员，充当职员
tick [tik] n. 滴答声，记号，勾号   v. 滴答地响，打钩
ventilation [ˌventi'leiʃ(ə)n] n. 通风，空气流通，通风设备

Technical Phrases

double-legged socket	两眼插头
electric shock	触电，电击
RS232 code	RS232 一种串行接口标准，有时指这种接口
VGA	模拟视频图像
NOTEBOOK	此处指笔记本电脑
portable computer	移动计算机，也笔记本电脑的一种说法
digital platform	数字平台

15.2 Reading materials

15.2.1 Operational instruction (selection)

Usually software operational manuals explain their operational processes in detail with graphic example, here are some examples which are selected from protel tutorial.

1. Schematic Layout- Adding Nets

On simple projects, it makes sense to run wires making connections, but on larger projects this often makes things very cluttered, difficult to make changes, and often results in errors due to wires crossing.

Draw a small wire extending from VSS (GND), and a small wire from the other pin on the pin header. Click the 'NET' icon in the toolbar (Fig 15.5) and place this text on both lines, labeling them both GND (as shown in Fig 15.6). This is the same exact thing as actually running wires between them. Protel knows that these two pins should be connected.

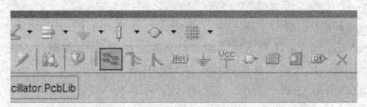

Fig 15.5　Click the "NET" icon in the toolbar

You should now have the following (Fig 15.6). Notice the blue dots on pin 12 and pin 11, the dots appear when more there is a connect of more than 2 areas. In the case of the blue dot on the left, this connects pin 12, pin 31, and GND all together.

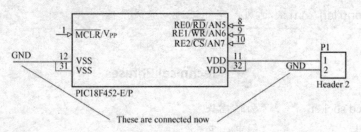

Fig 15.6　connect wit NETS

The usefulness of nets cannot be over emphasized here, it may seem like extra work, but DO IT!!

(1) Allows for cell design of layout, This makes design neat, orderly, easy to troubleshoot.

(2) Looks professional. (Fig 15.7)

(3) Major changes such as changing an 8 connection port from port A to B is simple to just

move the NETS, rather than rewire 8 different wires.

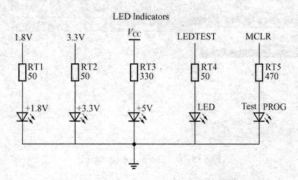

Fig 15.7   Schematic Layout- Using Nets

(4) Can create multiple schematic documents (advised), and link nets across schematics

2. Creating Custom Parts

Sometimes a part will be required and no library associated with it. For this, we need to make both a schematic and a PCB footprint for the part. Following are some steps of making a schmatic for a part.

(1) From the file menu (Fig 15.8), create a new schematic library, and a new PCB library. Save these libraries under a name you will remember.

(2) Select/Highlight the schematic library you just created and click 'SCH Library' in the bottom corner of the project menu (Fig 15.9).

Fig 15.8   file menu   Fig 15.9   Select/Highlight

(3) This shows information about the schematic library and a list of the parts inside the library. Click the ADD button to add a new part.

(4) Click on the place rectangle for our part (Fig 15.10). Keep in mind this is what we will see on our schematic.

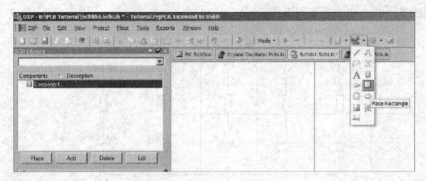

Fig 15.10   the place rectangle

15.2.2   Material of PGA202/203 (section)

Digitally Controlled Programmable-Gain
INSTRUMENTATION AMPLIFIER [Fig 15.11(a)]

The PGA202 is a monolithic instrumentation amplifier with digitally controlled gains of 1, 10, 100, and 1000. The PGA203 provides gains of 1, 2, 4, and 8. Both have TTL or CMOS-compatible inputs for easy microprocessor interface. Both have FET inputs and a new transconductance circuitry that keeps the bandwidth nearly constant with gain. Gain and offsets are laser trimmed to allow use without any external components. Both amplifiers are available in ceramic or plastic packages. The ceramic package is specified over the full industrial temperature range while the plastic package covers the commercial range.

A simplified diagram of the PGA202/203 is shown in Fig 15.11(b). The design consists of a digitally controlled, differential transconductance front end stage using precision FET buffers and the classical transimpedance output stage. Gain switching is accomplished with a novel

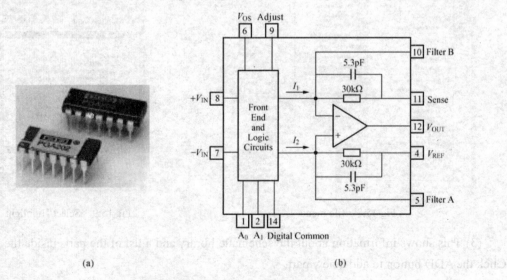

Fig 15.11   PGA202/203 and its diagram
(a) PGA202/203; (b) diagram of the PGA202/203

current steering technique that allows for fast settling when changing gains. The result is a high performance, programmable instrumentation amplifier with excellent speed and gain accuracy.

Fig 15.12 shows the proper connections for power supply and signal. The power supplies should be decoupled with 1mF tantalum capacitors placed as close to the amplifier as possible for maximum performance. To avoid gain and CMR errors introduced by the external components, you should connect the grounds as indicated. Any resistance in the sense line (pin 11) or the VREF line (pin 4) will lead to a gain error, so these lines should be kept as short as possible. To also maintain stability, avoid capacitance from the output to the input or the offset adjust pins.

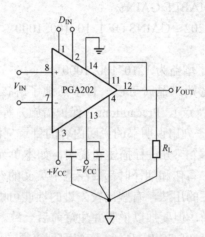

Fig 15.12　basic connection

15.3　Knowledge about translation（用户说明书）

使用说明书是随各种产品设备附带的书面材料，其形式有书本、小册子或散页之分，现在还有光盘（电子文件），主要看它所说明的设备大小与复杂程度而定，其内容包括设备的性能、安装、试验、维护等。

1. 说明书（用户手册）的特点

英语说明书在编写体裁等方面的特点是：

（1）文句简短、扼要。

（2）内容严格按照使用和维修时的先后顺序编排、划分类别章节，并给醒目的标题及不同的序号。

（3）做参考用的附图和附表比较多，用以辅助文字的说明，直观性强。

（4）重要的零部件和操作程序均用大写字体标出。

（5）在操作与测试过程的说明中，祈使句的出现率大，被动语态也很常见。

（6）常用到很多组合词，翻译时注意与实际结合。

2. 说明书的一般格式

(1) 序——说明书的最前面往往有个序,用来简洁强调本产品的优点或特点,有时也说明本产器使用前应特别注意的事项,如关于安全使用,电视机、录像机等的电源要求,对所接收信号的要求等,这一部分有时直接标明:本产品特点,注意事项等。

This VTR can be used with a power (mains) voltage of 100 to 110V, 115 to 127V, 200 to 220V or 230 to 250V.

这个录放机的电源适用范围为:100 到 110V,115 到 127V,200 到 220V or 230 到 250V。

DIGITALLY PROGRAMMABLE GAINS:
DECADE MODEL—PGA202,GAINS OF 1, 10, 100, 1000

数字可编程增益:
十进制芯片——PGA202,增益为 1, 10, 100, 1000

其主要用途是让用户在选购产品时可以一目了然地了解该产器的最主要特征及是否适用。还有常用的:Features(特点)、Precautions(保护措施)等。

(2) 目录(Contents)——列出说明书的说明内容的目录,便于查寻。

(3) 正文——对设备的各个部分进行描述和说明,如本单元课文及阅读材料所摘选的,一般是采用图文结合的方式,写作和翻译时都要注意中英文的对照,有时中英文的习惯表示方法并不一样,注意专业词汇的用法,有些比较先进的性能可能一时没有现成的中文对应,可以不译,也可以音译或意译,但要加以说明并注意前后一致。

Ceramic DIP, Plastic DIP or PLCC Package. 陶瓷双列直插,塑料双列直插或 PLCC(塑料有引线芯片载体,一种封装形式)封装。

如果有多项功能时,要注意逐项顺序介绍,重复步骤可写"参见……",文字要简洁,可以加上编号或项目符号,可以让使用者容易理解,方便操作。

正文可以采用章节结构,也可以采用项目结构或列表结构,视具体需要说明的内容而定。在使用英语大写字母缩写时,第一次必须有全文表示,使用专业术语最好也用通俗的语言甚至图片加以说明,要考虑到说明书(手册)的阅读对象并非一定是专业人员。

15.4　Exercises

1. Put the Phrases into English (将下列词组译成英语)
(1) 多媒体设备
(2) 电子窗帘
(3) 安装设备
(4) 触电
(5) 设备损坏
(6) 按下按钮

(7) 录像机信号

(8) 指示灯

(9) 遥控器

(10) 单击鼠标

2. Put the Phrases into Chinese (将下列词组译成中文)

(1) video presentation platform

(2) abide by the following proceeding

(3) working environment

(4) non-professional

(5) chemistry substance

(6) carry out finish class order

(7) light turns red

(8) uncontrollable button

(9) double click mouse

(10) delaying time

3. Sentence Translation (将下列句子译成中文)

(1) It's not convenient for the teacher to manipulate all these above.

(2) In order to guarantee the reliable operation of the equipments and safety of the staff, please abide by the following proceeding in installation, using and maintenance.

(3) Only after pressing the "CLASS BEGIN" can other button be controllable.

(4) Point the remote controller directly to IR receiving window and press corresponding buttons.

(5) Press the learning button within 10 seconds in order to make IR learning successful.

4. Translation (翻译)

把下列 PGA202/203 器芯片的特点和应用译成英语（有时用户说明书为了吸引大家的注意，用很简洁的方法列出产品的特点和应用，这里就是一个例子）。

FEATURES:

① DIGITALLY PROGRAMMABLE GAINS:

DECADE MODEL—PGA202

    GAINS OF 1, 10, 100, 1000

BINARY MODEL—PGA203

    GAINS OF 1, 2, 4, 8

② LOW BIAS CURRENT: 50pA max

③ FAST SETTLING: 2's to 0.01%

LOW NON-LINEARITY: 0.012% max

④ HIGH CMRR: 80dB min

⑤ NEW TRANSCONDUCTANCE CIRCUITRY

⑥ LOW COST

APPLICATIONS:
① DATA ACQUISITION SYSTEMS
② AUTO-RANGING CIRCUITS
③ DYNAMIC RANGE EXPANSION
④ REMOTE INSTRUMENTATION
⑤ TEST EQUIPMENT

15.5　课文参考译文

随着现代教育的发展，很多教室中都添加了各种多媒体设备，例如、多媒体电脑、录像机、VCD、投影机、视频演示平台、电子窗帘和功率放大器等。教师要去操作所有这些设备有时不太方便。为了解决这个问题，人们开发了一种教育控制系统。通过控制计算机上的多媒体设备或者控制面板，上课时控制所有这些设备就十分方便了。

本手册介绍 PC-950/PC3900 教学控制系统的使用方法。

15.5.1　安全操作指导

为了保证设备可靠工作和人身安全，在安装、使用和维护过程中请按照如下步骤进行操作：

（1）安装设备时，确认输入电源的接地线是完好的，不要用两眼插座，并确认输入电源电压与主机上所标的（额定电压）是一致的。

（2）内部有 110V/220V 高电压元件，没经过允许不要打开机箱，以防触电。

（3）不要把系统设备放在太冷或太热的地方。

（4）因电源工作时产生热量，工作环境应该维持通风良好，以防止过热引起的损坏。

（5）在长时间不用或潮湿季节切断电源开关。

（6）在进行下列操作前，确认交流电源已切断（交流电源插头已拔出）。

① 拔掉或重新装上设备的任何部件。

② 拔掉或重新连接设备的任何插口。

（7）非专业人员或未经允许，请不要打开设备的机箱，不要自己维修，以免出事故或进一步损坏设备。

（8）不要让任何化学物质或液体溅入设备。

15.5.2　PC-950 前面板介绍

PC-950 教学控制系统的前面板如图 15.1 所示。

1. 控制钮

（1）上课或下课

① 按"上课",电子屏放下,投影机通电,7s 后投影机开始工作。如果再次按"上课"就没有任何作用了。

② 按"下课",系统按序在一定时间内完成结束工作,电子屏开始上升(收好),投影机切断 RS232 通信接口。这时(下课后),只有"上课"按钮有效。

注意:只有按下了"上课"按钮后,其他的按钮才受控。

(2) 投影视频选择

① 按"DVD/TV/VCR",可把模拟信号从 DVD 通过 RS232 串行口传送给投影机(当切换成 DVD 时"DVD/TV/VCR"指示灯转成红色)。第二次按这个按钮,就把模拟信号从电视送给投影机(当切换成 TV 时"DVD/TV/VCR"指示灯转成绿色)。第三次按这个按钮,把 VCR 的模拟信号送给投影机(当切换成 VCR 时"DVD/TV/VCR"指示灯转成橙色)。

② 按"PC/NOTEBOOK/VIS",可把 VGA(视频图形阵列)和音频信号从计算机通过 RS232 串行口传送给投影机(当切换成计算机时"PC/NOTEBOOK/VIS"指示灯转成红色)。第二次按这个按钮,就把 VGA 和音频信号从笔记本电脑送给投影机(当切换成笔记本电脑时"PC/NOTEBOOK/VIS"指示灯转成绿色)。第三次按这个按钮,把数字平台的 VGA 信号送给投影机(当切换成数字平台时"PC/NOTEBOOK/VIS"指示灯转成橙色)。

(3) 音量控制

① VOL+ 输出音量增加

② VOL− 输出音量减少

(4) 投影控制按钮

① ON——按下"ON"按钮,从 IR 和 RS232 送出打开投影机的信号。

② OFF——按下"OFF"按钮,从 IR 和 RS232 送出关闭投影机的信号。

③ VIDEO——按下"VIDEO"按钮,IR 信号和 RS232 送出视频信号给投影机进行显示。

④ VGA——按下"VGA"按钮,从 IR 信号和从 RS232 送出 VGA 信号给投影机进行显示。

(5) IR 学习

包括按钮"功能 1,功能 2,ON,OFF,VIDEO,VGA",步骤如下:

① 按下"IR LEARING"键持续 2s,就转入 IR LEARING 模式,这时其他按钮都处于不受控制状态。"IR LEARING"指示灯最多亮 10s,如果没接收到任何 IR 代码灯就会灭了。

② 把遥控器直接对准 IR 接受窗口,按下相应的按钮,则 6 个 IR 学习按钮接收到 IR 代码的信号后闪烁,如果 10s 内接收不到任何 IR 代码就会退出学习模式。

③ 在 10s 内按下学习按钮使 IR 学习成功,然后它的灯闪三次后灭掉。

2. 移动电脑(笔记本电脑)VGA 信号输入口

连接移动电脑(笔记本电脑)的 VGA 接口,标准的 15 针 HDF 接口。

3. 网络接口

与后板上的互联网口直接连接。

4. 电子锁

只有一个特殊的钥匙才可以打开这个锁，它等于"上课"和"下课"按钮的功能，当锁上后所有的按钮都不受控。

5. USB

与后板上的 USB 端口直接相连。

6. 模拟信号输入

从外部设备输入模拟信号，默认设置后，模拟信号通过扩音器或投影机输出。

7. 话筒

连接到话筒。

15.5.3 软件功能

图 15.2 为教育控制系统软件的主界面。

1. 开始上课

对"开始上课"单击左键就开始上课，击右键则打开下列窗（图 15.3），可以设置是否允许"投影"和"投影屏幕"工作。选择需要的打钩，然后单击"下载"确认。

选完"投影机动作"和"投影屏幕"后就开始送出 RS232 串行信号和 IR 信号使投影和投影屏幕工作。

2. 结束上课

对"结束上课"单击左键结束上课，单击右键则打开下面窗口（图 15.4），可以设置是否允许"投影"和"投影屏"关闭，选择需要的打钩，然后单击"下载"确认。

还可以设置延时时间：

设置延时结束上课的时间范围：0～60min

设置投影机工作关闭延时时间：2～6min

然后，确认。

通过选取"投影"和"投影屏幕"，系统会通过 IR 和 RS232 端口传送信号给投影机和投影屏幕使之关闭。

15.6 阅读材料参考译文

15.6.1 操作手册（节选）

软件操作手册一般都用图例来详细解释操作过程，这里是选自 protel 手册上的一些例子。

1. 原理图设计——加上网络

在简单的项目（原理图设计）中，可以直接用线进行连接，但在较大的项目中，如果直接画连接线会非常乱，很难修改而且在线与线交叉时容易出错。

从 VSS（GND）端画一条短线，在另一个引脚的端画一条短线。在工具条（图 15.5）中单击"网络"图标，然后在这两条短线上都放这个文本框，都标为 GND（如图 15.6 所示）。这样做和把这两条线连接起来的效果完全相同。Protel 知道这两个引脚是连接的。

现在应该如图 15.6 一样，注意在 12 脚和 11 脚的蓝点，这些点表示有两个以上的地方需要连接在一起。在左边的蓝点情况中，引脚 12、引脚 31 和 GND 都要连接在一起。

这里强调用网络的方法表示器件之间的连接，看起来建立网络好像是多余的，实际上非常有用。其有以下的好处：

（1）可以分块设计原理图，这样使得图整洁，有序，容易发现并改正错误。

（2）使设计图看起来比较专业（图 15.7）。

（3）如果要作大量的变化例如要把连接到端口 A 的 8 根线改为连接到端口 B，只要简单地把"NETS"网络标记移动一下，而不需要重新画这 8 根的连接线。

（4）可以（建议）创建多张电路原理图，用网络连接把多张图连接在一起。

2. 创建一个自己的元件

有时需要一个元件，但元件库里没有，我们就需要建立一个元件，包括它的原理图符号和 PCB 引脚。建立一个元件的原理图符号的步骤如下例所示。

（1）从文件菜单（图 15.8），创建一新的原理图（符号）库和一个新的 PCB（符号）库，把这些库保存起来，起个容易记的名字。

（2）选取刚建立的原理图库（使之成为高亮表示已选取）。单击底部项目菜单上的"SCH 库"按钮（图 15.9）。

（3）这时显示关于这个原理图库的信息和库里的元件的列表，单击"ADD"按钮加一个新的元件。

（4）单击放矩形框来画我们的元件（图 15.10），记住，这就是我们等会儿要在原理图中看到的样子。

15.6.2 PGA202/203 资料（摘选）

数控增益可调——仪用放大器 [图 15.11（a）]

PGA202 是单片仪用放大电路，由数字控制其放大倍数在 1、10、100、1000 四级可调。PGA203 的放大倍数则为 1、2、4、8 四级可调。这两个芯片都有与 TTL 或 CMOS 匹配的输入，很容易与微处理器连接。两个芯片都有 FET（场效应晶体管）输入端和一个新的反馈电路，使得输出在带宽范围内几乎保持不变。增益和偏置是激光微调的，可以不接任何外部器件。两个放大器都有陶瓷或塑料两种封装产品。陶瓷封装可用于整个工业温度范围（比较大），而塑料封装只能用在消费品级温度范围。

PGA202/203 的框图如图 15.11（b）所示，放大器由数字控制电路，采用精密的 EFT 缓冲器反馈电路的前端级和一个典型的跨阻输出电路组成。其中增益（放大倍数调节）开关是用一个新型电流换向技术的电路来实现的，当需要改变放大器的增益（放大倍数）时可以快速调节。总之，这是一个高性能，可通过编程改变放大倍数，响应速度快和放大倍数精确的仪用放大器。

图15.12给出电源和信号的连接方法。电源的耦合电容用1mF钽电容，耦合电容尽可能地靠近放大器的引脚，可以使放大器的性能达到最佳。为了避免由于外部器件引入的增益和CMR［Common Mode Rejection（ratio）共模抑制（比）］误差，耦合电容要如图15.12接地。在感应引脚（11 脚）或参考引脚（引脚 4）端的引线电阻可能会引起增益误差，所以这些引线应该尽可能地短。为了维持稳定性，输出到输入或输出到偏置调整引脚之间不要接电容。

参 考 文 献

[1] Ken Martin. 数字集成电路设计（英文版）. 北京：电子工业出版社，2002.
[2] Bergen.A.R. 电力系统分析（英文版）. 2版. 北京：机械工业出版社，2005.
[3] Allan R. Hambley. 电子技术分析（英文版）. 北京：电子工业出版社，2005.
[4] Leonard S.Bobrow. 线性电路分析基础（英文版）. 2版. 北京：电子工业出版社，2002.
[5] 朱一纶. 电子技术专业英语. 2版. 北京：电子工业出版社，2006.

参 考 资 料

[1] http://en.wikipedia.org/wiki/Engineering
[2] www.howstuffworks.com
[3] www.elect-spec.com
[4] Introductory tutorial, Exploring Protel 99SE（软件附带的手册）
[5] www.mikroelectronika.co.yu
[6] www.altium.com